DYE PL

Dye
Plants
AND DYEING

JOHN & MARGARET
CANNON

ILLUSTRATED BY
GRETEL
DALBY-QUENET

Timber Press
in association with
The Royal Botanic Gardens, Kew

First published in North America in 1994 by
Timber Press, Inc.
The Haseltine Building
133 S.W. Second Avenue, Suite 450
Portland, Oregon 97204, USA

This edition 2003

ISBN 0-88192-572-1

A CIP catalog record for this book is available
from the Library of Congress.

Designed by Pauline Harrison
Cover designed by Dorothy Moir

Printed in Hong Kong

Contents

Foreword

Since early times, primitive man was attracted by colour and began by using colours to paint his body, varying from the use of woad by the Celtic peoples to the use of annatto by the Amazon Indians. Both of these plants are described in this welcome book. Many of the body paints were later used as dyes for textiles.

This book, which covers a fascinating array of natural plant dyes, is an important new contribution to the literature because it provides accurate botanical information about each of the species treated and comes out of the authors' own experiments with plant dyes. Here you will find how to use as dyes parts of many familiar plants, from the wallflower to the yew, from nettles to elderberries. Some of the dyes described come from further afield, such as the Brazil wood, which gave that country its name because of the quantity of this valuable dye exported in the early days of colonization. The book is full of interesting historical information as dyes have been important since the early days of the human race. I feel sure you will enjoy browsing through the beautiful paintings of Gretel Dalby-Quenet.

I am particularly happy to see the authors' concern for the conservation of plant species that provide the dyes. So often in the past people have been profligate with natural material; the Brazil wood almost became extinct in Brazil through overuse. It is, moreover, good to learn that synthetics have not altogether taken over from natural dyes. I know that *Dye Plants and Dyeing* will encourage you to experiment and to see for yourselves the wonderful colours that nature has produced.

GHILLEAN T. PRANCE
Director
Royal Botanic Gardens, Kew

Introduction

At some time in the distant past, early man found that plant pigments could make textiles more colourful and enliven a somewhat drab existence. This discovery may not have been as important as those concerned with the development of agriculture, or the knowledge of plant drugs to combat sickness, but it played a significant part in the development of human relationships with the natural world. The first attempts to gather this information, during classical and medieval times, led to efforts to classify the many different types of plants and their uses. This early work culminated in the eighteenth century with the great Swedish botanist Linnaeus creating the basis of modern systematic botany.

As professional botanists, we have found satisfaction in combining the use of scientific data with the production of beautiful and subtle colours; this book is the result of an attempt to bring together these two interests. Many craft books presently available provide excellent instructions on practical dyeing with natural materials, but are generally lacking in botanical information on the plants involved. The interest and enjoyment of many craft dyers can be greatly enhanced by knowing more about the materials they are using. This particularly relates to the classic dyestuffs, primarily tropical in origin, available to us through the craft suppliers, and which, prior to the development of synthetic dyes by the chemical industry, provided many of the colours that brightened the lives of our ancestors. Most dyers using, for example, brazil wood or fustic, know next to nothing about the trees from which the materials originate.

The drawings of the plants we discuss are based on the long tradition of botanical illustration, which has always aimed to combine aesthetic excellence with scientific understanding and accuracy. We have been fortunate in having the very close cooperation of our friend Gretel Dalby-Quenet, whose illustrations greatly enhance the information in this book. The coloured illustrations are all drawn from living material, while most of the sepia drawings are based on museum specimens.

Many craft workers have an instinctive concern for the conservation of the world's resources and we have stressed the many species that can easily be grown in our gardens, and the need for restraint in collecting wild material. We also hope that importers of dyestuffs will be mindful of their moral obligation to ensure that their suppliers are not pillaging dwindling resources in the developing world to meet the needs of affluent craft workers in the more developed countries. We have included information that makes this book of particular interest to dyers in western Europe, North America, Australia and New Zealand.

Unless otherwise stated, all the results reported in the text come from our own experiments, and the colour samples with the drawings have been matched with these. We have kept the plant descriptions as simple as possible and for those unfamiliar with botanical terms we have provided a short glossary. We have also included some coverage of the biochemistry of plant pigments, since the dyestuffs produced by plants are no less 'chemical' than those used by modern industry. Natural substances react in similar ways to those produced synthetically and some dyes, such as indigo, can now be produced artificially. We are particularly grateful to Dorothee Goldman who has guided our own somewhat faltering steps in this very technical field and has critically reviewed these parts of the manuscript. A tribute is merited by Sir Edward Bancroft, whose classical works on dye pigments published last century are veritable mines of information.

Sincere thanks to friends, botanists and dyers, too numerous to mention individually, for their help and interest. Our particular thanks to Anne Hecht who read the text in early draft; her comments have enhanced the book's content.

We very much hope that dyers will find their work stimulated by a better understanding of the plant materials they are using and that botanists will find greater interest in the practical uses of these plants.

JOHN & MARGARET CANNON
Barn Croft, Rodmell, Lewes, East Sussex BN7 3HF

Beginnings in dyeing

EQUIPMENT

You do not need to spend much on equipment to be successful at dyeing. Basic necessities are a water source, a container, a heat source and dye materials. At the other extreme, however, a 'dye-house' with stainless-steel utensils, and other expensive equipment can be established although, for most craft dyers, the kitchen with some everyday utensils and a few specially purchased items is a realistic compromise.

WATER SOURCE A naturally soft water supply is best, but many dyers using hard water can still produce excellent colours. Hard tap water can be softened with a small amount of 'Calgon', but an excess of this makes the wool hard and causes yellowing. Rain-water is often slightly acid and, if drained from the roof, can contain iron from the gutters which will modify your results, as can the algae, fungi and bacteria which grow in the collecting butts.

DYE POTS Ideally these should be kept solely for dyeing and never used for cooking. Stainless-steel buckets or large pans are best, but large, unchipped enamel casserole dishes or pans may also be used. Containers should be large enough to hold at least five litres of water for every hundred grams of wool (one gallon to four ounces). Aluminium preserving pans may also be used when mordanting with alum, but can modify the colour of many dyes. Similarly, copper, tin and iron containers will have an effect on the resulting shades. Glass vessels are ideal in a microwave oven.

HEAT SOURCES An adjustable heat source is advisable, either electricity or gas, but bear in mind that spills must be cleaned up at once to avoid permanent staining. Microwaves can be used for test runs and for softening fruit.

STIRRERS These can be wood or glass. Many plastics are not suitable because they bend in boiling liquids.

MEASURING JUGS Glass is preferable. Plastic should not be used for coloured liquids or chemicals, particularly if the jugs are to be used for any other purpose. Plastic buckets and washing-up bowls are excellent for rinsing the dyed fibres.

SCALES A 'Gourmet' food balance is ideal for weighing small quantities; for larger amounts domestic scales can be used. In both cases, the scale pan must be covered with cling film, or the weighing done in a small container, and never directly on to the pan surface.

THERMOMETER A kitchen or laboratory thermometer is needed for indigo and madder dyeing.

GLOVES Rubber or plastic gloves are essential when dyeing with indigo or woad, or handling some dyestuffs such as annatto and alkanet.

STORAGE JARS A range of plastic pots and glass jars with lids will be needed for storing chemicals and dyestuffs.

DRYING Wool is best squeezed thoroughly or given a very short time in a spin-dryer before hanging on a line or rack in the shade.

FIBRES

Natural dyes can be used for many kinds of fibres, but this book is mainly concerned with wool, which is most frequently used by amateur dyers. Other fibres, such as flax and cotton, require different preparation and mordanting, and some craft dyers have obtained very successful results with silk. Artificial fibres take up the pigments to a greater or lesser extent, but are not considered here. Wool from sheep and other animals can be dyed spun or unspun, but in either case it must be carefully prepared beforehand. Natural oils, such as lanolin, which help to protect the animals, form a covering over the fibres and must be removed before dyes or mordants are applied. Soil particles and plant fragments also prevent even take-up of the pigments, as do traces of dung and urine that may be present. The process of cleaning and degreasing is known as scouring. Wool should be soaked in cold water for several hours and then squeezed thoroughly. A warm bath should be prepared using either soap flakes or a mild detergent; the wool is then gently squeezed in it and then rinsed. If the wool is very dirty this process may have to be repeated several times. When the wool is clean, a spin for a very short time in a spin-dryer will remove most of the water. Wool must always be wetted before treatment with mordants or dyes; skeins should be soaked for at least an hour in cool water before use and the addition of a little detergent hastens the wetting process.

Spun wool is formed into skeins which are loosely tied in several places, with a figure-of-eight twist of cotton, to prevent the skeins from becoming entangled during the dyeing. Cotton is used because it takes up dyes differently from wool and is easily found and removed when the dyed skein is to be wound into a ball.

NATURAL DYESTUFFS

Natural dyes come not only from flowering plants, but also fungi, lichens, insects, shellfish and various earths, though only flowering plants are included in this book. Dye materials can be gathered at most times of the year: young leaves and flowers in spring, mature leaves and flowers in summer, fruit in autumn and bark and roots in winter. All may be used in their season.

FLOWERS Many flowers give excellent dyes; a rough guide is to use twice the weight of fresh flowers to the weight of the fibre. Some are best used fresh, but many are just as good after drying on trays in a warm airing cupboard until thoroughly dehydrated. Equal quantities of dried flowers to fibres are used. Delicate flowers are often boiled together in the same bath as the fibre, so that their pigments are kept at a high temperature for as short a time as possible. Some species produce much the same colour from flowers, young shoots, leaves and stems. When these parts are used together, they are referred to as 'flowering tops'. The brightest dyes are usually produced from freshly collected material, although petals from faded or even nearly dead plants can produce good results.

FRUITS Fruits are easy to gather and prepare. Berries with thin skins, such as raspberries, blackberries and elderberries, need only brief cooking and crushing to extract the juice. A short time in the microwave speeds up the process. Harder fruit, such as sloes or privet, may need soaking for twenty-four hours before cooking and should be crushed as far as possible before soaking - a large pestle and mortar is ideal, but a block of wood in a bucket can be used as a substitute. About equal weights of fruit and fibre are needed, but sometimes it may be necessary to increase the proportion of fruit to fibre. Soft fruit from the freezer can be used; blackberries and raspberries stored in this way have produced satisfactory results, but blackcurrants and damsons were disappointing. Immature walnuts in their husks can also be used and can be gathered as they fall, placed in a bucket of water and kept in a cool place. Unlike many other fruit, they rot very slowly; some kept for over two years gave near-perfect results when crushed and boiled in the usual way.

LEAVES Leaves can be gathered at most times of the year, but those from deciduous plants may produce very variable results according to their age. About double the amount of leaf to fibre is needed. Tough, leathery leaves, such as those of the evergreen holly, ivy and cherry laurel, should be torn into pieces and soaked for a few days before use. Cherry laurel must be used with caution because it emits cyanide which is toxic, causing dizziness if inhaled, and, since the yellow-greens produced are not particularly attractive, it is better avoided.

BARKS Barks give a wide range of colours in the red-brown-yellow range. Small twigs can be peeled and the soft bark soaked and boiled. Allow equal weights of bark to fibre. Mature bark can be collected from fallen trees and branches. The outer corky bark should be discarded and the soft inner layers retained. Oaks give a pleasing range of buffs and browns, alder gives browns and yellows (grey to black with iron), beech gives orangey-pinks, while birch provides soft pinks and browns. Mature bark needs to be chopped up and soaked for several days before use. Many barks contain tannins and produce colours without further mordanting and are thus classed as substantive dyes.

ROOTS Roots should only be gathered when there is no risk of damaging the plant concerned or when major changes are being made in the garden. For example, digging away a weedy bank in autumn gave us an excellent batch of tormentil roots, which, after treatment, yielded soft apricots and browns. Fallen trees are another source and the retrieval of some London plane roots from road works in a street gave beautiful pinks and browns. Roots and rhizomes grown specifically for dyeing, such as madder, alkanet and turmeric are dealt with elsewhere in this book, but, in general, roots should be chopped, soaked (for several days if woody) and used with equal amounts of fibre, though up to double the weight of dyestuff to fibre may sometimes be needed.

The amounts of dyestuff to fibre given here are not very precise because so many variables are involved in natural dyeing. A quick test run can be made by doubling the amount of dyestuff and using only a few grams of wool, for example, 5g (⅙oz) of wool to 20g (¾oz) fresh flowers. If there is a strong colour after only a few minutes of simmering, it might be better to halve the weight of flowers. Colours seem to fade more rapidly if the fibres are in contact with the pigments for only a short time, and it is easier to regulate the depth of colour during a longer heating period. While it is very difficult to lighten the result, a paler tone can always be darkened by longer simmering or adding more dyestuff.

PREPARATION OF THE DYE BATH

The estimated amount of dyestuff is placed in the chosen container together with the amount of water required – at least 5 litres (1¼ US gallons) per 100g (3½oz) of wool. The pan is then slowly brought up to simmering point and kept at this temperature until no more dye appears to be coming out of the dyestuff. Allow to cool and then carefully strain off the solid material. The bath is then ready for the wetted wool to be added and for the dyeing to begin.

MORDANTS

Most natural dyes are fixed by chemical bonds within the structure of the fibres. Some are known as *substantive* (or direct or non-mordant) dyes and are absorbed without further chemical treatment – walnuts and onion skins are good examples. The majority of dyes, however, need additional chemical treatments to promote their absorption and to prevent fading and 'bleeding' of the colours. These are known collectively as *adjective* (or mordant) dyes. The chemicals used are called *mordants* and these fix the dye to the wool fibres. They are also used to alter and enhance the colours. Mordants work by forming chemical bonds between the dye molecules and the proteins of the wool fibres. A few dyes, such as indigo, do not penetrate the fibres and only become loosely attached to the outside, which is why jeans become blotchy with washing and wear.

Mordants can be naturally occurring substances, for example, urine, wood ash, plant galls, tannins and crabapple juice, or pure chemical compounds with metallic ions. These are commonly salts of aluminium, chromium, iron, tin and copper, though others are occasionally used. They are usually bought as pure substances from specialist retailers, or are introduced by using copper or aluminium dye vessels. A few rusty nails or an old wirewool scouring pad can be used as a source of iron. However, a number of common mordants are poisonous and must be handled with care. Commonsense practices such as not cooking, smoking or eating while handling them will promote safety, as will rigorous procedures for storage and labelling.

Mordants can be used before, during or after the dyeing process.

PRE-MORDANTING This is the most commonly used method, and it mordants the fibre before dyeing. The mordant is dissolved and added to a bath containing ample water. The fibre is added and the whole brought slowly to the boil over a period of about three-quarters of an hour and simmered for a further quarter of an hour in the case of fine wools, or longer when coarse

13

wools are used. It is then allowed to cool and the fibres are moved gently during this period to ensure the even take-up of the mordant. The wool can be used immediately, the fibres being drained and gently squeezed, or put in a spin-dryer for a few seconds to remove excess moisture; or they can be kept moist for a few days in a sealed plastic bag; or stored for future use after thorough drying. Chrome-mordanted wool should be used at once or after storage for a few days in complete darkness as it is very sensitive to light (some residual mordant effect remains after storage for several months in the dark).

SIMULTANEOUS MORDANTING This gives faster results by cutting out mordanting as a separate stage. It has the advantage of less fibre handling, although the colours may not be as permanent as with separate mordanting. The mordant is first dissolved and then added to the bath containing the prepared dye. The wetted-out fibre is then added. The whole is brought slowly to nearly boiling and the fibre is gently moved with a stirring rod from time to time. Simmering continues until either all the dye is taken up or the required shade is reached, remembering that the wool will become several shades lighter during rinsing and washing. The fibre can either be left to cool in the liquid, or removed and rinsed in very hot water, followed by subsequent rinses of progressively cooler water. The fibre should then be gently washed with soap or mild detergent until no further colour is lost. After thorough rinsing, wool skeins are often weighted to straighten the threads during drying. Fleece is dried flat on a towel and finished cloth either laid flat or wound on a slatted roller.

AFTER-MORDANTING Mordanting after dyeing is also possible, and tin and iron are often used in this way, after pre-mordanting or simultaneous mordanting with another substance. Tin has the effect of 'blooming', i.e. it enhances and brightens, while iron 'saddens' or dulls colours. They are added to the dyebath for the final five to ten minutes of simmering, the fibres having been lifted out while the dissolved salts are thoroughly mixed into the dye liquor.

The action of mordants and other additions does not depend on their dilution; the fibres take up these substances in relation to their own weight. All you need is sufficient water to cover the wool and allow it plenty of room to move freely. It is quite satisfactory to dye several skeins premordanted with different salts in the same dyebath, because once combined with the fibres, the metallic ions do not react with one another. It is necessary in this case to label each skein to distinguish it from the others after dyeing. A system of knots is often used, a short length of a different fibre is attached to the skein, with one knot for alum, two for chrome, three for copper, four for tin and so on. In order to distinguish skeins treated with acids, alkalis or other assistants, buttons of different shapes can be attached. A careful note of the method used should be made.

THE USE OF COMMON MORDANTS

ALUM [Al] Potassium aluminium sulphate (potash alum) or ammonium aluminium sulphate (ammonia alum).

To mordant 100g (3½oz) wool:
    18g (⅝oz) alum
    6g (⅕oz) cream of tartar

Place the chemicals in a container holding about 5 litres (1¼ US gallons) of water, stir well until the crystals are dissolved, then add the clean and thoroughly wetted-out wool. Bring slowly to boil over about an hour, simmer gently for a further hour and allow to cool in the liquor. The wool should be gently turned with a rod during the process to make sure it is evenly mordanted. It can be used immediately, kept damp in a plastic bag for a few days, or dried for long-term storage.

Gill Dalby (1985) recommends that less alum should be utilised. The amount used can be reduced by using 8% alum and 7% cream of tartar using the above method. The wool should be thoroughly rinsed and left in a sealed plastic bag or airtight container for three to five days. This is said to produce brighter, more light-fast colours.

CHROME [Cr] Potassium dichromate
This should be stored in a dark jar away from light. There are disadvantages in using chrome because it is poisonous, light-sensitive and some people are allergic to it. However, it often produces extremely attractive colours, some of which are more permanent than those produced with alum, and it also leaves the wool feeling silky and lustrous.

To mordant 100g (3½oz) wool:
   3g (⅛oz) chrome
   6g (¼oz) cream of tartar

The method is the same as for alum, but the wool should be used as soon as possible after mordanting or stored in the dark for a few days only.

Industrial dyers use formic acid as an assistant when dyeing with chrome. Gill Dalby in her book *Natural Dyes, fast or fugitive* (1985), suggests that craft dyers should use it also, as by using a 2% solution of formic acid with 1% potassium dichromate less chrome needed. The improved absorption of the mordant means that a lot less chrome is left in the mordanting bath, making its disposal more environmentally friendly. Concentrated formic acid is exceedingly dangerous to handle. NEVER try to dilute your own.

The use of chrome as a mordant is now somewhat controversial, some dyers prefer not to use it at all, whilst others find the colours it give to fibres too exciting to neglect.

COPPER [CU] Copper sulphate
Copper often gives darker shades to reds and browns and may give greenish shades where alum gives yellows. The colours are sometimes more light-fast than those resulting from alum.

To mordant 100g (3½oz) wool:
   12g (⁷⁄₁₆oz) copper sulphate
    6g (¼oz) cream of tartar (optional)
or 2g (¹⁄₁₆oz) copper sulphate
   100ml (3½oz) vinegar

Vinegar gives an unpleasant smell, but the consequent reduction in the amount of copper used is ecologically desirable.

TIN [Sn] Stannous chloride
This can be used as a pre-mordant, but it is most frequently used near the end of the dyeing process to give added brightness, particularly to wools pre-mordanted with alum. Tin can precipitate or decolour pigments in the dyebath, and can make wool very harsh and brittle, so it should be used with great caution. A satisfactory quantity to use is: 3 to 4g (⅛oz) tin, with 4g (⅙oz) oxalic acid to 100g (3½oz) wool. The stannous chloride should be dissolved in a small amount of water and added to the dyebath for the last five to ten minutes of boiling. Be sure to take the fibres out of the bath while the tin solution is added, otherwise there may be uneven take-up of the mordant.

IRON [Fe] Ferrous sulphate
This is used in a similar way to tin. Iron often saddens or dulls colours and an excess leaves the wool harsh and brittle. Dissolve 5g (³⁄₁₆oz) of ferrous sulphate crystals in water for each 100g (3½oz) of wool. Add to the dyebath for the last five to ten minutes. Many people prefer to use less ferrous sulphate, finding that 2g per 100g of wool is sufficient.

RHUBARB LEAVES I

In her book *Wild Colour* (1999) Jenny Dean recommends the use of oxalic acid as a mordant. The pure chemical is very toxic and difficult to obtain, rhubarb leaves which contain considerable quantities of this acid can be substituted. They should be simmered for an hour in enough water to cover them. The solution is cooled and used as a mordant, but beware, the leaves are very poisonous, use rubber gloves and make sure to dispose of the remains carefully.

DISPOSAL OF RESIDUES

Never use more of the chemicals than is advised in the mordanting recipes; it is not only wasteful but may harm the texture of the fibres. Ferrous sulphate, for instance, can seriously damage wool; an excess of it in the dyebath causes wool to become lifeless and brittle. The disposal of the residues from mordanting baths sometimes presents problems; it is essential to aim for the least possible amount to be left after mordanting. A sensible article by Jocelyn Banks appeared the *Journal for weavers, spinners and dyers* No 73 (March 1995), explaining why merely pouring everything away down the drain is not considered acceptable. Dyers with large amounts of chemicals for disposal should consult this reference.

Most craft dyers only use small amounts of chemicals at a time. Alum is used in gardens as an acidifier for plants such as rhododendrons, camellias and heathers. There should be very little in any spent mordant bath, especially if cream of tartar has been used. Dilute the spent bath with ample amounts of water and sprinkle on the soil, especially round acid-loving plants. Iron is used by plants in the manufacture of chlorophyll (their green pigment). Any chlorotic (yellowing) plants could benefit from the remains of an iron mordant bath, otherwise it can be diluted and safely poured on the ground. Copper is used by gardeners as a fungicide and occasionally on peaty soils as a conditioner. Well-diluted and acidified with vinegar, it may be poured onto soil quite safely away from streams or wells, as can the spent tin mordant bath well diluted with water. Chrome is the most difficult mordant for disposal. Every effort should be made to completely exhaust the mordant bath, in the first place by very careful weighing of the both the amount placed in the bath and the fibre to be dyed. Underestimate the amount to be used if in any doubt, and do not dispose of the waste if any yellow colour is left in the bath. If any colour remains add some waste wool, and simmer until the bath is quite colourless. After much dilution, it can then be poured away.

Proper disposal techniques will vary according to the chemical and locale. Before disposing of any chemical, read the directions for disposal if these are provided, or seek advice from your supplier. In the United States, suppliers and retailers must provide, on request, a Material Safety Data Sheet, with information on proper and lawful handling and disposal, for every chemical. Those intending to operate semi-professionally, who will have relatively large amounts of waste for disposal, must contact the local sewerage company or chemical waste removal companies.

ASSISTANTS

Supplementary substances used in dyeing and mordanting are often termed 'assistants'. The following list includes those most frequently used.

CREAM OF TARTAR [Potassium hydrogen tartrate]
Often used with alum and sometimes with other mordants. It combines with the metallic salts of mordants to give compounds which attach more easily to the fibres; it also brightens colours and helps to keep the wool soft.

The chemical known in cooking as cream of tartar is sodium pyrophosphate; this is less effective as an assistant.

OXALIC ACID
A poisonous substance found naturally in rhubarb, wood sorrel and other plants. Most often used to assist the absorption of tin.

TANNIC ACID
Found in many plant species, especially in the barks of trees and shrubs. Needed in cotton dyeing.

ACETIC ACID
Often used to brighten colours when using berries. Commercial vinegar contains 4 to 6 per cent acetic acid and is a convenient source. Garments made with colours strongly modified by acids should always, after washing, be rinsed in water containing a little vinegar because the colours may be altered by neutral or alkaline tap water.

FORMIC ACID
Used to assist the absorption of chromium in mordanting. It is chemically related to acetic acid, but the unpleasant fumes it gives off make it more difficult to

use. It must be obtained from laboratory suppliers and is unlikely to be available in small quantities.

AMMONIA

A few drops of household ammonia alters the pH (the measure of acidity or alkalinity) towards the alkaline ranges. This markedly changes the colours of some dyes and it may almost decolourise a few that are particularly ammonia-sensitive. Colours that are greatly influenced by pH should not be mixed, as subsequent washings may alter them differentially (*see also* under acetic acid).

GLAUBER'S SALTS [Sodium sulphate]

Used to even colours, they act by slowing the speed of pigment take-up. The salts are added to the dyebath during the dyeing process, to help fix the colours and prevent bleeding. Another use is to 'level' colours dyed in different batches; two skeins of similar shades but dyed in different batches can be closely matched if they are simmered together in a solution of the salts.

COMMON SALT [Sodium chloride]

Sometimes used in rinsing to prevent excessive bleeding.

WASHING SODA [Sodium carbonate]

Used in the same way as ammonia to alter the pH. Too much will harden the wool and an excess can completely dissolve it.

PLANT PIGMENTS

The most important pigments that are used for producing colours on wool and other fibres can be classified as follows:

ANTHRACENES

These include several pigments found in the Madder family such as *alizazin, mungistin,* and *purpurin; emodin* (from Persian berries) and *polygonin* (from Japanese knotweed). The insect dyes cochineal, kermes and lac are also included here. Napthoquinones are related compounds and include *juglone* (from walnuts) and

alkanin (from dyer's alkanet). *Hypericin* (from St John's wort) is a condensed quinone.

CAROTENOIDS

Some well-known examples are vitamin A and pigments found in egg yolk. Those used for dyeing include *lutein* (from nettles, French marigolds and many other plants); *bixin* (from annatto) and *crocin* (from saffron).

FLAVONOIDS

(a) ANTHOCYANINS produce the scarlets, reds, violets and blues of nearly all flower petals. They include *malvidin* and *cyanidin* both of which are found in several plant families.

(b) FLAVANOLS AND FLAVONES are responsible for the yellow colours in the flowers of many plants. When they occur with anthocyanins, the yellows may be masked and the petals show more of the blue colours. Flavonol dyes tend to fade somewhat in strong light. The most common pigments are *quercetin, kaempferol* and *myricetin,* all found in several plant families – and in large quantities in instant tea powder. *Fisetin* occurs in fustic, nettles and other plants and *morin* is found in fustic, Osage-orange and other related species. Flavones often produce colours that are more permanent, but rather paler than those of the flavonols; *apigenin* (common in the Daisy family) and *luteolin* (common in the Daisy and Pea families and also present in weld) are examples.

(C) MINOR FLAVONOIDS include chalcones and aurones, which are often found together, and isoflavones. All three are often present together with other flavonoids and may modify the colours produced. Chalcones include *coreopsidoside* and *mareoside* (common in the Daisy family), and the aurones include *sulphuroside* (in young fustic and some members of the Daisy family). Isoflavones often produce strong, fairly permanent colours; they are found mainly in the Pea family and include *genistein. Osajin* and *pomiferin* are found in Osage-orange.

TANNINS

These are found in small quantities throughout most plant tissues, but are concentrated in bark and in damaged tissues such as galls and wounds. They are formed from flavonoids, particularly anthocyanins, during the breakdown of tissues, and are used commercially in tanning leather. Chemically divided into hydrolysable and condensed tannins, the former give rather yellower colours to wool than the latter, and include *gallotannins* (found in oak galls and young fustic) and *ellagitannins* (found in several families including the sumac and water-lilies). Condensed tannins tend to give more reddish tones and are found in numerous plant groups, particularly in oak and willow bark, cutch and water-lily rhizomes.

UNRELATED PIGMENTS

Mention is included in the text of a number of pigments that are not related to any of the groups classified above. These include *indigotin* (from indigo, woad and other species), the diaroylmethane *curcumin* (from turmeric), pterocarpans (from sanderswood) and the neoflavonoid *berberine* (from barberry and mahonia).

SCIENTIFIC PLANT NAMES

In this book we give the recommended English name (Dony, Jury & Perring 1986) for all species that occur in the wild in the British Isles, together with some common synonyms and vernacular names used in the United States. The names of species which originate outside these areas have been selected from appropriate Floras for the countries concerned. These are followed by the scientific names which use an internationally agreed Latin format, and may be essential to distinguish two or more quite different plants which have the same common name. The family to which the plant belongs helps to provide a framework of reference that can lead to more botanical information. An appendix gives a list of all the plants referred to in this book. Set out in families, the list gives the currently accepted botanical name and author, and any common synonyms.

The first part of a Latin scientific name indicates

the genus to which the plant belongs and is somewhat like the surname of a human family. The second name is the specific epithet and is rather like a given or Christian name. People are frequently mystified by the abbreviated personal names, such as L., R.Br and Hook.f., that follow the Latin names of plants. These indicate the botanists responsible for the recognition of the plant as a hitherto unknown species, and they are used by scientists as a further safeguard against confusion in the application of the name. Frequently, a little knowledge of Latin can give a clue as to the plant's properties. For example, *Rubia* is derived from *rubra* which means red and refers to the colour given by roots of many members of this genus. *Tinctoria* refers to dyeing properties, so *Alkanna tinctoria* suggests that this species is a good source of dyestuff - which is indeed the case.

The use of the vernacular name 'alkanet' has caused great confusion among dyers, because it is applied to plants with greatly differing dye properties. The true dyer's alkanet is *Alkanna tinctoria*, with woody roots which produce a strong purple dye. Before 1824, it was included in a related genus *Anchusa*. The common 'anchusa' of gardens is also known as 'alkanet', although it is more correctly called green alkanet or *Pentaglottis sempervirens*. This species has rather succulent roots which, at best, only produce a dirty brown colour. Further complications come from the dyers' use of the term 'redwoods', which is applied to species from several families of flowering plants that produce red or purple dyes. To Californians and botanists in general, redwoods are the magnificent coniferous trees from the northern California coast, which include the tallest trees known to man. These can also be used for dyeing, but give entirely different colours. These two examples demonstrate clearly how confusing common names can be without the back-up of formal scientific nomenclature.

It is worth taking considerable trouble to make sure that the plants you use for dyeing are correctly named, so that you can make proper records and discuss your results with others. One last example will serve to reinforce this important idea. It is relatively easy to mistake knapweed (*Centaurea nigra*) for saw-wort (*Serratula tinctoria*). The former gives a yellow that is quite different from that produced by the latter.

CONSERVATION

Conserving the world's diversity and the rational use of renewable resources cannot be stressed too highly. This responsible view fits in well with an appreciation of materials and good craftsmanship. For this reason, emphasis is placed on plants that can easily be grown in our gardens, minimising demands on plants in the wild. People who sell tropical dyes from Third World countries will often tell you the materials have been supplied without harming the dye plant's fragile environment. The problem is, you, as a buyer, do not know if this is true. Consequently, all craft workers should be careful to buy their requirements from reputable dealers.

For the same reason, lichens have been excluded, although they are a very traditional source of dyes in the British Isles. For some years, it has been the policy of The Natural History Museum, London, to recommend that lichens should no longer be used. They grow very slowly, cannot be cultivated despite some suppliers' claims to the contrary, and in recent times have suffered from the effects of air pollution, to which they are exceptionally vulnerable. Furthermore, lichens are not easily identified and long hours of preparation can give disappointing results if the wrong species has been gathered. The colours they produce can be spectacular, but permanence is very poor. A leaflet is available from The Natural History Museum which suggests alternative sources for lichen colours.

While the use of garden materials is stressed, it is not, of course, intended that the moderate and considerate use of materials from the countryside should be excluded. Prohibiting the use of abundant species such as elderberries, nettles and brambles on the grounds of conservation would be ridiculous and counterproductive. However, some of the other species included here, such as the saw-wort, are never very abundant and are best left untouched, as in any case seed is available from specialist suppliers. The UK *Conservation of Wild*

Creatures and *Wild Flowers Act 1975* and the *Wild Life and Countryside Act 1981*, and the US *Endangered Species Protection Act* and other local and state regulations, provide considerable protection for native flora, including a total ban on picking many very rare species. However, the ordinary craftsperson is unlikely to encounter these and our thoughts should therefore be directed towards the abundance of common species that enrich our countryside. Keep in mind though, that it is an offence to uproot any wild plants without the permission of the landowner.

Craft dyers should become proficient in identifying wild plants and should never collect any material unless certain of its identity. Even then, natural populations should be left strictly alone unless it is obvious that collection will not jeopardise their future.

Criteria such as 'Do not collect more than 10 per cent of a plant population' have been suggested but, in reality, the natural environment is so complex and variable as to make such well-meaning generalisations almost meaningless. The only guideline must be: if you have the slightest doubt in your mind, don't collect the material and leave the plants for others to enjoy.

KEY TO SYMBOLS USED ON THE PLATES

The mordants and assistants used to produce the colour samples shown in the illustrations are indicated by the following symbols:

Alum Al
Chrome Cr
Copper Cu
Tin Sn
Iron Fe
Acid medium (ac)
Alkaline medium (alk)

Conversion tables

grams	g/oz	ounces	centimetres	cm/in	inches
28.35	1	0.04	2.54	1	0.40
56.70	2	0.07	5.08	2	0.80
85.05	3	0.11	7.62	3	1.20
113.40	4	0.14	10.16	4	1.60
141.75	5	0.18	12.70	5	2.00
170.10	6	0.21	15.24	6	2.40
198.45	7	0.25	17.78	7	2.80
226.80	8	0.28	20.32	8	3.20
255.15	9	0.32	22.86	9	3.50
283.50	10	0.35	25.40	10	3.90
567.00	20	0.71	50.80	20	7.90
850.50	30	1.06	76.20	30	11.80
1134.00	40	1.41	101.60	40	15.80
1417.50	50	1.76	127.00	50	19.70
1701.00	60	2.12	152.40	60	23.60
1984.50	70	2.47	177.08	70	27.60
2268.00	80	2.82	203.20	80	31.50
2551.50	90	3.17	228.60	90	35.40
2835.00	100	3.53	254.00	100	39.40

5ml = 1 US teaspoon
15ml = 1 US tablespoon
30ml = 1 ounce
1 litre = 1.76 pints (Imperial)
2.13 pints (US)
4.6 litres (approx) = 1 gallon (Imperial)
3.7 litres (approx) = 1 gallon (US)

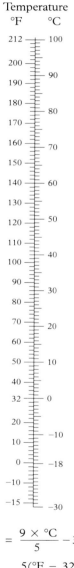

Temperature °F °C

$$°F = \frac{9 \times °C}{5} - 32$$

$$°C = \frac{5(°F - 32)}{9}$$

Notes for teachers

The process of dyeing wool, or any other fibre, can be just as rewarding for children as it is for adults. By using plants from the garden, or plant material such as tea leaves and onion skins, children can be shown how colours are easily produced. A classroom experiment will also give them a greater appreciation of plants, particularly if they use materials they see every day around the home or school. To reinforce this, it is important for each child to take home some dyed wool. Unbleached wool gives the best colours and it can sometimes be bought from shops that sell knitting wool on 'cones'. If it is unavailable, white (bleached) wool can be substituted. Thick wool gives more solid colours than thin, but takes longer to dye, but whichever is chosen, it must be thoroughly soaked before use and a few drops of detergent in the water hastens this process. If several sessions are planned, the children can premordant the fibre. However, if only a short time is available, it is better to use substantive dyes without mordants, or wool premordanted by the teacher. Additives such as vinegar or bicarbonate of soda are quite safe and give interesting colour changes with various dyes.

A first demonstration could be with onion skins, boiled the night before the class; wool, cloth, or scraps of fleece, if available, can be soaked before the class begins. A very strong extract of onion skins gives results very quickly on mordanted wool. Dilution of the dye bath will produce a range of paler shades, while adding vinegar, bicarbonate of soda, or old wire-wool pads will give further colour changes.

Skeins can be simmered in the dye bath without any mordant, then a small amount of alum added to give deep orange in a strong bath and brilliant yellow in a weaker bath.

The dye fibres should be rinsed out and washed quickly in a little detergent, rinsed again and hung over a line, or laid flat on plastic bin liners to dry. The maximum amount of skeined wool to the quantity of liquid in the dye bath is 60g (2⅛oz) for every five litres (1¼ US gallons). Children can bring dyes from home, such as instant tea, or teas from different countries, instant coffee, turmeric, walnut husks or leaves, and acorns, all of which will give good results without mordants. If mordanted wool is to be used, tomatoes, red cabbage, carrot tops, young bracken shoots, and various berries and flowers can be tried. Younger children should be discouraged from bringing blue flowers as they usually produce dull greeny-yellows and even the brightest blue delphinium flowers give disappointing results. Beetroot will probably be suggested, but this only gives drab and dull fawns; it could, however, be used to illustrate that not all brightly coloured substances give bright colours to wool. Do not allow the children to handle the wet dyed fibres with their bare hands; use plastic or rubber gloves.

The dyed fibres can be used in a variety of ways, for example, in collages, embroideries and simple woven mats. Notes should be kept recording the plants and methods used. The demands of conservation and the proper use of natural resources should also be stressed.

The Plants

Annatto (Lipstick plant)

Bixa orellana (Annatto family – Bixaceae)

Annatto is native to South America. It is widely cultivated commercially as a food colouring, and is also used as an ornamental or hedge plant. Other common names include achiote, arnotto, rocou, urucu and otter. In Germany it is called 'Orleans', after confusing the name of the French city (Orléans) with that of Francisco de Orellana, the famous Spanish explorer in whose honour it was named. It is a bush or small tree of 2 to 3m (6–10ft), with somewhat elongated heart-shaped leaves. The branching inflorescences bear pink five-petalled flowers which produce showy red capsules, covered with soft spines or hairs. When mature, the capsules open to expose up to fifty yellow-red seeds, in two double rows which are embedded in a soft, bright red pulp. It is this pulp that contains the dye.

The dye was used by many Amazon tribes as a body paint and had religious as well as magic significance. It was known to the Mayas and Aztecs of Mexico and to the Incas of Peru. In the West Indies, the pulp was mixed with lemon juice and used as a body paint as recently as the late eighteenth century. Annatto was exported to Europe from Guyana, Jamaica and the Antilles before the introduction of synthetic dyes. Cardon (1990) notes that it was imported into Portugal in the sixteenth century as 'terra oriana', and was being grown for commercial use in India in 1787. It was used with a number of other dyes such as brazil wood and weld to counteract its tendency to fade. Bancroft (1813) recorded the dye as being used principally on silk and sometimes on cotton; the fresh pulp produced a more permanent dye on cotton than dried pulp. The permanence of the colour improved if the pulp was separated from the seeds by washing, precipitated with lemon juice or vinegar, and then used in the normal way. It is widely used today as a natural food colouring known as E.160(b) in Europe, particularly for margarines and milk and cheese products, because it is tasteless and harmless. The name 'lipstick plant' refers to its use in the cosmetic industry; it is used in many hair lotions and body oils.

Annatto needs tropical or sub-tropical conditions to thrive and it is easy to cultivate in such conditions. It can readily be grown from seed and will bear fruit after about two years. Dried seeds, together with their pulp, can be bought in Europe from dye suppliers. In the Caribbean and parts of the United States, the seeds with their pulp can be bought in grocery shops and ethnic food stores and are used to colour scrambled eggs and other food. In areas where it is cultivated, fresh pulp is separated from the seeds by crushing and fermenting, after which the seeds are removed and the pulp is made into cakes or paste. In the West Indies, the cakes are wrapped in banana leaves and sold as 'flag annatto'.

The home dyer is usually advised to soak the dried seeds in water or vinegar for at least a week, stirring or shaking every day. If the reconstituted pulp is not soft enough to use, it may be boiled for about an hour in vinegar. The method advised by Krohn (1980) is 50g (1¾oz) dried seeds, 350ml (¾ US pint) vinegar, 4 litres (8½ US pints) water and 200g (7oz) salt to 150g (5¼oz) wool. Seeds should be placed in a muslin bag, heated to almost boiling in the vinegar and water, then the salt dissolved and added to the dye bath. The wetted and warmed wool is simmered in the dye bath for an hour, allowed to cool, rinsed thoroughly and then dried in the shade. Krohn notes that pyridine is used commercially to extract the dye colour. It is soluble in oil, but this may be difficult for the home dyer to remove from the wool. Cardon (1990) recommends boiling the pulp in washing soda for use on cotton; this process would not, of course, be suitable for use on wool.

The dye substances present are tannins and flavones, together with a number of carotenoids, of which *bixin* and *nor-bixin* are the most important. The yellow-to-red dyes produced by annatto are substantive, but they can also be modified by the use of mordants. In alkaline solution, the *nor-bixin* is particularly active. None of the colours produced are especially light-fast; some fade quite rapidly and fading or bleeding may also occur after washing with soap.

Cu

Al

Baptisia

Baptisia australis (Pea family – Leguminosae)

Baptisias are native to the central and eastern United States, and there are about seventeen species, most occurring in light woodland. The name comes from classical Greek and refers to its dyeing property. *Baptisia australis* is a moderately robust perennial herb, 1 to 1.5m (3–4½ft) high, with glabrous, green leaflets of 5 to 7cm (2–2½ins), arranged in groups of three. The flowers have almost navy blue petals and are borne on an elongated spike. The mature pods are black and inflated and the seeds rattle inside the dry pods, which explains the American common name 'rattle flower'.

Baptisia is a most ornamental addition to the flower border and the pods enhance dried flower arrangements. It can be grown in almost any soil and tolerates a limy soil very well. The plant dies down in the late autumn, although the fruiting stems may persist for much of the winter. These should be cut back as near the base as possible before the new shoots appear. Baptisia can be bought at many nurseries, and mature plants can be successfully divided in late autumn. The seeds do not appear to germinate very readily, but scoring the hard outer coat with a file before planting may give some success.

Leaves and flowering stems, or flowers only, can be used in dyeing. The flowers need to be boiled for just a short time with the fibres to give a good colour; the leaves also release their colour fairly rapidly. The flowers without the leaves give an intense, almost luminous green with alum, an orangey-brown with chrome and an olive-green with copper. Unmordanted wool gives a paler version of the alum colour. The leaves give similar colours; 'saddening' the colour with iron provides an attractive dark green. Fading, tested by exposing the dyed fibres to a week of almost continuous sunshine in midsummer, showed a marked yellowing with the alum- and non-mordanted samples, but had very little effect on the others.

Most of the dye substances fall within the flavonoid group, flavones including *spartein*, *luteolin*, *apigenin* and several flavonols are recorded; Baptisia probably also contains *indican*.

False indigo (*B.tinctoria*) is a smaller plant. The leaves are not as bluish as those of *B.australis* and the inflorescences have fewer, yellow flowers. A native of the southeastern United States, spreading northwards as far as Ontario, Canada, it is found in light woodland, and may have been grown commercially in the United States before the invention of coal-tar synthetic dyes. 'Indigo' (including *Indigofera* species) was cultivated in the hope of replacing the true dyestuff, which was expensive and imported from India, but there is no firm evidence that Baptisia was involved. However, the implications of its common name, and that it contains *indican* (which produces a blue dye after suitable treatment), makes it highly probable that it was used for this purpose. Like *B.australis*, it is also rich in flavonoids including *pseudobaptigenin* and *orobol*, which could also have been a reason for its cultivation.

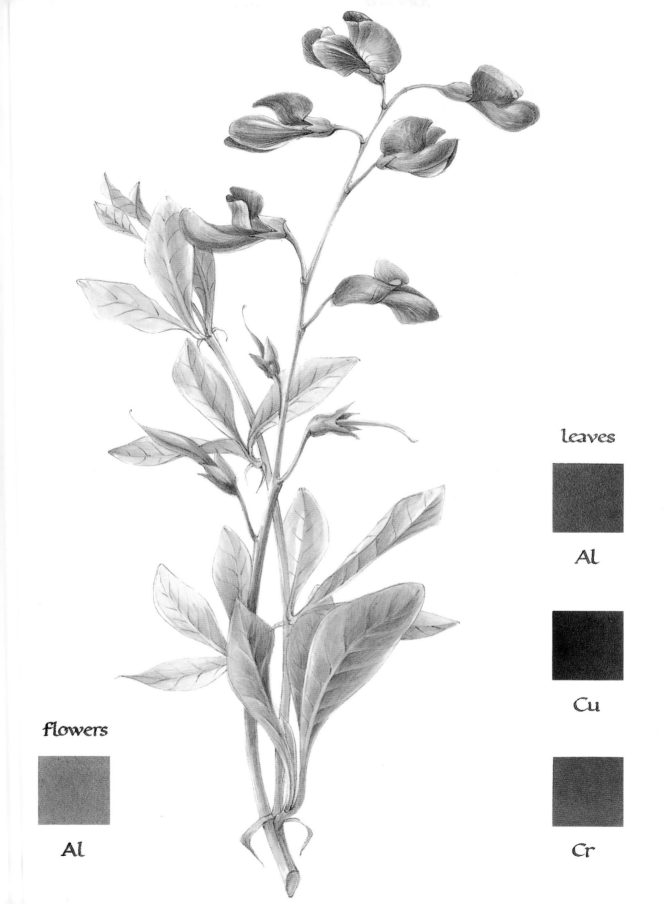

leaves

Al

Cu

flowers

Al

Cr

Black oak

Quercus velutina (Beech family – Fagaceae)

The black oak is a native of the eastern and mid-western states of the United States. It is also known as the yellow or American oak and it produces a dyestuff known commercially as quercitron. It is a moderate to fast-growing tree of up to 25m (85ft), occurring on rocky hillsides and sandy ridges. The bark is brown to black on the trunk and dark reddish-brown on the branchlets. The leaves are 10 to 20cm (4 to 8ins) long, with up to seven deep lobes, dark green and shiny on top, paler beneath. The pointed acorns are up to 2cm (¾in.) long and the cup, which covers about half the nut, has dark brown hairy scales. The tree is now grown commercially for lumber, but is of major historic importance as a source of yellow dye.

The dye bark was named 'quercitron' by Edward Bancroft, who brought it back from America thinking it could be a substitute for weld. It quickly became popular in Europe following its introduction in the 1770s. Huge amounts were imported until the beginning of this century, when it was largely superseded by synthetic dyes. It was planted extensively in France, where large quantities of acorns were sent to the Trianon and other royal establishments. Bancroft was granted a five-year monopoly by Parliament in 1770 for the importation of the bark into Britain, where the dye was eagerly accepted by dyers, particularly in the cotton industry. The monopoly was of great financial importance to him, and in 1775 he made a profit of £5000. He attempted to have his patent extended in 1776. Although the House of Commons approved the bill, it failed in the House of Lords, where the members included a number whose wealth was dependent on the cotton industry. This was ironical because the failed bill would have forced Bancroft to sell at a fixed (and lower) price for seven years, whereas the unrestricted importation of the bark resulted in quadrupled prices.

The bark contains tannins and was used extensively in the leather industry. The famous chemist Perkin discovered that two separate dye compounds were present in quercitron: these are *quercetin* which gives golden-yellow with alum, brownish-yellow with chrome, lemon-yellow with tin and olive with iron; and *quercetagetin* which gives orange-red with alum, orange-brown with chrome, bright orange with tin and olive-brown with iron. Industrial extracts include *flavine*, which some claim is still unsurpassed by synthetic dyestuffs for producing a brilliant, vivid yellow on wool mordanted with alum and tin. *Quercetin* has recently been reported in the press as the active constituent of red wine which apparently has a beneficial effect in warding off heart disease.

Unfortunately for home dyers, it is rarely possible to obtain quercitron extract, and anyone wishing to experiment with this fascinating yellow dye must extract it themselves from black oak bark. It is not difficult to get a good colour from young branches of about 2.5cm (1in.) in diameter. The outermost layers of the bark contain a yellow dye which is easily dissolved by boiling; it produces a good yellow colour with alum. The thicker inner layer also contains large amounts of pigments; these give oranges and tans, possibly because of the amounts of tannin present. Our own experiments suggest that the outer layers of the bark contained more *quercetin* than *quercetagetin*, and the inner layers the reverse, but it is possible that differing amounts of tannins are responsible for the results. Bancroft (1813) advocates the use of glue or gelatine to give purer yellows.

There are two native oaks in Britain, both of which also occur over much of Europe. The pedunculate or common oak (*Quercus robur*) has leaves with little lobes at the base next to the leaf stalk and acorn stalks of 2 to 8cm (1 to 3ins); while the sessile or durmast oak (*Q.petraea*) has leaves without the basal lobes and fruit stalks of up to 1cm (½in.) Both species give similar dyes: young leaves picked in June and soaked for four days give shades ranging from buffs and beiges to mid-brown. Bark from a tree which fell in February gave pale beiges and golden-browns. Adding iron gave dark greyish-browns with alum and chrome and a solid dark brown with copper. Acorns give pale browns and tans.

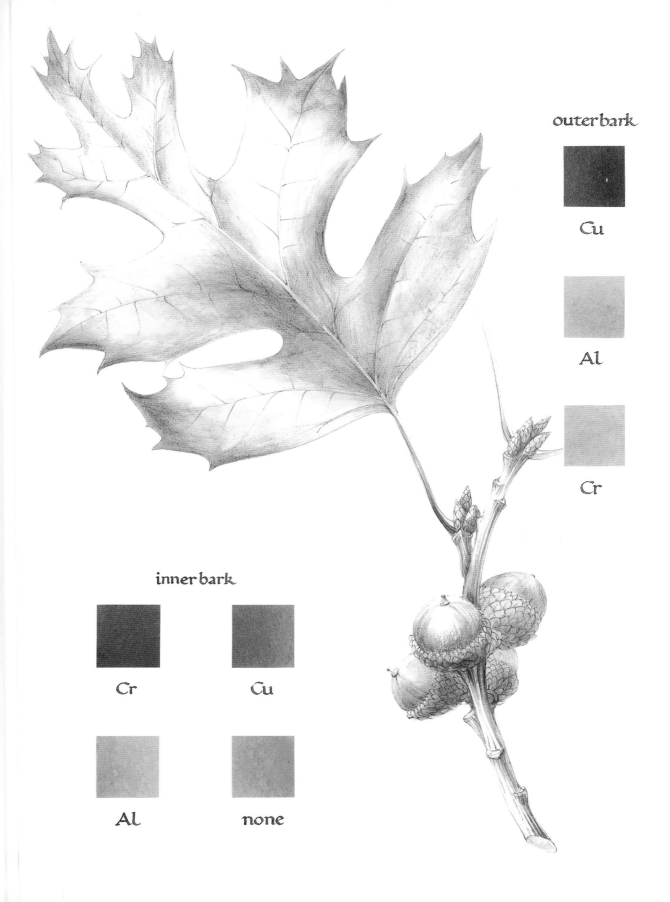

outer bark

Cu

Al

Cr

inner bark

Cr

Cu

Al

Black walnut

Juglans nigra (Walnut family – Juglandaceae)

There are about twenty species of walnut which are widely distributed in temperate and tropical regions. Black walnut is a native of the central and eastern United States, where it is found on well-drained soils at low altitudes, most often occurring in mixed forests and rarely forming pure stands. The roots secrete a substance which leaches into the soil and prevents the growth of seedlings (and those of other species) near the established tree. It is widely planted in the United States and other temperate areas of the world, including Europe, and matures moderately quickly. The timber is highly regarded for cabinet making, while the nuts are used in confectionery and ice cream.

Walnut dyes are of great historic importance. In the first century AD, Pliny records their use to keep hair from turning white. His recipe included the use of walnut shells (probably husks) boiled with oil, ashes, lead and earthworms. In the *Ladies Dictionary* of 1694, walnut husks were used in hair dyes to make grey hair black. The recipe states: 'Hair, to render it black, take the bark of an oak root, green husks of walnut, three ounces of each, and the deepest and oldest red wine a pint. Boil them well bruised to the consumption of half a pint, strain out the juice and add oil of myrtle a pound and a half. Set them six days in the sun in a leaden mortar, stirring them well, anointing the hair will turn any coloured hair and black as jet in often doing.' The use of a lead mortar must have been highly dangerous as the acid in the wine would have reacted with it. During the making of the famous Gobelins tapestries, husks were covered with water and left to ferment in a warm place for at least two years before use. Another species of *Juglans*, the butternut or white walnut (*J.cinerea*), was used by early settlers in the United States and during the Civil War when other sources of yellow or yellow-orange dyes were unavailable. So many Confederate soldiers were clothed in homespuns coloured with this species that the nickname of 'butternuts' was given to them. The related pecan nuts (*Carya illinoensis*) are now familiar in British shops and are imported from the United States and Australia. The hard outer shells, which are naturally pale brown, are frequently artificially dyed red for the British market.

Black walnuts grow to about 40m (130ft). The bark is dark brown to black, each leaf has 9 to 15 leaflets which are slightly toothed at the edge and hairy underneath. The fruit have thick, slightly hairy, husks and the hard shells are not easy to split. Young trees can be bought at many nurseries. They should not be planted in areas susceptible to late frosts and, as they grow into very large trees, are not suitable for small gardens.

The parts most often used for dyeing are the leaves and fruit husks, but the bark, catkins and heartwood are also used. Leaves, fresh or dried, should be soaked for at least twenty-four hours before use. All parts of the tree contain a substantive dye, so it is not necessary to mordant wool to produce a strong colour. However, mordanting does produce a further range of shades, particularly with chrome, copper and iron. Husks are easiest to separate from the nuts while still fresh, and should be handled with rubber or plastic gloves, otherwise the hands will be badly stained. The husks can be left in a bucket of water for many months, or may be dried slowly for long-term storage.

All parts of the tree give various shades of browns and yellows. The colours are very permanent, except for pale shades with an alum mordant, which may yellow a little in sunlight. The bark removed from two-year-old branches is said to give a puce colour to wool mordanted with bismuth and tin, or brown-violet if given a very long simmering. The dye pot is said to smell of wallflowers (Buc'hoz 1800).

The common walnut (*Juglans regia*) is a much slower growing tree, planted mainly for its timber and nuts. It can be distinguished from the black walnut by its grey bark. The leaves have 7 to 9 untoothed leaflets, the husks are smooth, and the nuts are easy to split. The dye compounds are the same in both species, being derivatives of napthoquinones, of which *juglone* is the most important. This pigment is present in all parts of the plant and, when purified, gives a red-orange dye which can be modified by the presence of tannins and various flavonoids.

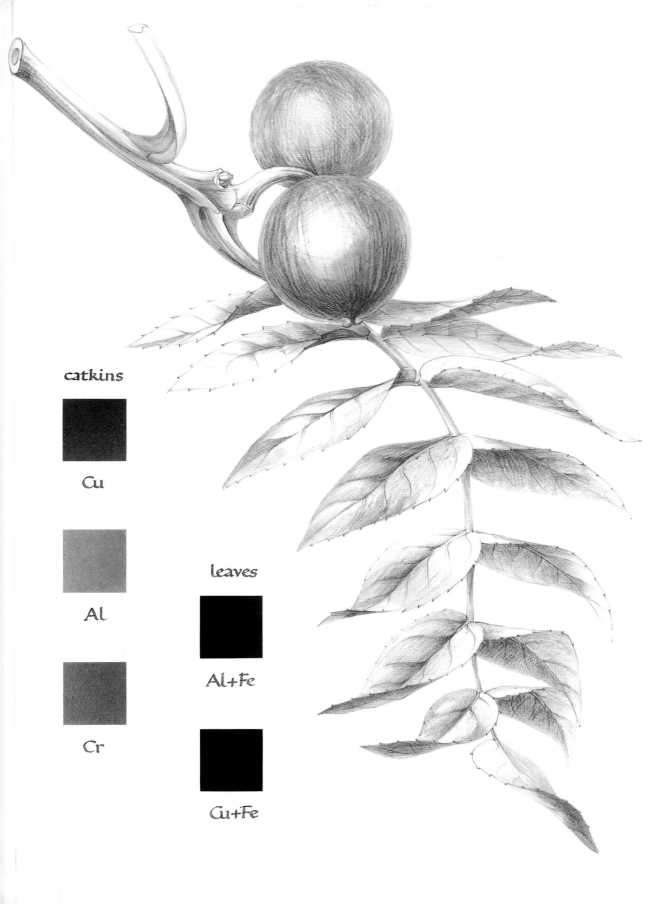

catkins

Cu

Al

Cr

leaves

Al+Fe

Cu+Fe

Bloodroot

Sanguinaria canadensis (Poppy family – Papaveraceae)

Bloodroot is a native of eastern and central North America, ranging from Nova Scotia to Manitoba in Canada, and south to Oklahoma, Alabama and Florida in the United States. It is usually found in moist woodland habitats and is a perennial with a thick, horizontal rhizome, which exudes a copious and very poisonous red sap when damaged. The plants die back in the autumn and, in the following spring, produce several shoots, each consisting of a flower and one basal leaf. The leaves are stalked, palmate, and heart-shaped at the base, with 5 to 9 deep irregularly shaped lobes, green above and bluish beneath. In April and May, the flowers, which are borne on leafless stalks, have 8 to 12 white petals and a boss of numerous golden stamens in the centre. The flowers open only in full sunlight and die after one or two days. The leaves continue to enlarge after the petals fall and may eventually reach 20 to 25cm (8–10ins) across. The seeds are wrinkled, glossy brown, and in the wild are dispersed by ants.

Bloodroot was used by various tribes of Indians in Canada and the United States, for dyeing the skin as a war paint when mixed with bear's grease, and historians refer to both yellow and red colours. The Ojibwa tribe are said to have used bloodroot to dye porcupine quills red; it is also stated that it was used to paint imitation blood on tomahawks. Young girls of the tribe are said to have used it for rouge. They made strong decoctions of the dye which they kept during the winter months when fresh roots were unavailable. Bancroft (1813) notes that bloodroot produced a bright orange on silk that had been mordanted with tin. It has also been used to treat diseases of the respiratory tract, and even to treat cancers, despite its poisonous nature. A moderate dose is said to produce vomiting and a large one may be fatal.

Bloodroot is available from specialist nurseries and some garden centres; a double form, in which the fertile parts of the flower are replaced by petals, is more easily obtained than the natural wild form described above. It has the advantage of a longer flowering time for individual flowers, but is not as beautiful and does not set seed.

The roots should not be disturbed after growth recommences in the spring, and large clumps should only be divided while dormant in the winter. Fresh seeds should be sown as soon as they are ripe in a mixture of loam and peat. Roots should be planted in a deep pocket of loam and peat or compost in a shady position, preferably under deciduous trees, as the flowers only open completely in full sunlight. Bloodroot can be grown in many soil types as long as plenty of humus is present.

The roots, which become orange on drying, are available from dye suppliers, and should be chopped and pounded before use. Simultaneous mordanting is sometimes recommended and the roots must be tied in a muslin bag and placed in the dye bath at the same time as the wool. The colours we produced from what appeared to be rather old material were dingy fawns after extraction with water, but the same roots soaked in alcohol gave much stronger colours. Unmordanted wool became a mid-orange colour, alum and tin gave tans, chrome a strong orange-brown and copper mid-brown. The dye pigments include the isoquinoline alkaloid *protopine* and many other alkaloids are also present.

alcoholic extracts

Cr   Sn   none   Al

Bramble (Blackberry)

Rubus fruticosus (Rose family – Rosaceae)

There are a very large number of European 'micro-species' of brambles, between 350 and 400 in Britain alone. They are very difficult to distinguish from one another and for practical purposes are often treated as one large aggregate species. Brambles vary in habit, leaf shape, petal colour and shape, prickles and many other characters, not least fruit flavour. This suggests that the microspecies may have varying dye properties, although it is unlikely that any differences would be significant.

Brambles can be grown in any soil and plants can be bought from garden centres and planted at any season of the year. Old wood should be cut back each autumn and the young shoots trained in convenient positions for gathering the fruit the following year. A note of caution: the branches have a habit of bending and rooting at the tips, so care should be taken, otherwise you could end up with an unwelcome bramble thicket. The berries can also be gathered from hedge-rows and wasteland.

Several parts of the plant can be used. Young shoots can be gathered throughout spring and early summer, but they should be collected before the wood hardens and while the prickles are still soft. Leaves can be collected at any time during the growing season but, as the tannins increase in late summer, they are, at this time, most useful for producing dark browns and blacks. Roots can be used, some giving orange, others greenish colours, but this may depend on the state of maturity. Fruit should be gathered when ripe and may be frozen for use during the winter.

The normal mordants, alum, chrome and copper can be used and the colours are considerably altered by the addition of iron, tin and acid or alkali. Leaves and young shoots give varying shades of yellows and buffs, which can be 'saddened' by iron to give attractive grey and brown shades; a larger amount of iron gives a particularly pleasing black with copper. Cut the leaves, shoots and roots into small pieces and boil in the usual way. The resulting colours are reasonably light-fast, but paler shades may brown a little in very strong sunlight. Fruit can produce reds, greens or purple. For these, boil ripe berries in a little water until quite soft, then strain through a moistened jelly bag. Avoid straining through a sieve because this leaves small particles of pulp on the wool fibres, making it very difficult to get an even shade. Adding vinegar or other acid to the dyebath gives pale violet and pinkish-purple shades with unmordanted wool, while a small amount of alkali produces blue-greys and slates. A very small amount of tin results in a beautiful rich violet. The colour-fastness depends partly on the depth of tone that has been produced; paler shades seem to fade more or become more tinged with brown than darker shades. Adding sugar, or even flour, to the dyebath is thought by some to enhance colour-fastness, on the principle that it is often impossible to remove a blackberry pie stain from clothing – but this may be just an old wives' tale. However, the proteins in flour may have some effect, although its use can cause uneven dyeing of the wool.

The pigments present in the berries are antho-cyanins, in particular *cyanidin* (known in the European food industry as E.163(a)), *malvidin* (E.163(c)) and *chrysanthemin* and related compounds which are very sensitive to changes in pH; tannins are also present. The leaves and stems are extremely rich in tannins.

Other species of *Rubus* may be used for dyeing. Raspberries (*R.idaeus*) can be used for red-purples, while the leaves and those of other species, such as *R.tricolor* and *R.allegheniensis* give similar colours to those produced by brambles and can be used in the same way.

berries

Sn (ac) Al+Sn (ac)

Cr (alk) none (ac)

shoots

Cu+Fe Cu Cr Al

Brazil wood

Caesalpinia species (Pea family – Leguminosae)

Caesalpinia sappan, a tree native to the Malay Peninsula and the Indonesian Islands, was the original 'brazil wood'; the name brazil means the colour of red-hot coals. It is a small, thorny tree with yellow flowers, and pods 8 to 10cm (3¼ to 4ins) long by 3 to 4cm (1¼ to 1½ins) wide.

The use of sappan wood is of great antiquity. It was exported from India to China as early as 900 BC (Cardon, 1990) and the Arabs were certainly using it and re-exporting it to Europe by early medieval times, as an act of the Scottish Parliament during the reign of David I (1084–1153) confirms (Grierson, 1986). We do not know when the name was altered to brazil, but it was imported into Provence under that name in the thirteenth century. Brazil wood was well known to the Portuguese who first colonized South America in about 1500, where they found large numbers of similar trees growing and hence named the country Brazil. New World brazil woods were soon imported into Europe where they quickly became important dyestuffs. By the 1830s, brazil wood had been almost completely superseded in England by camwood (*Baphia nitida*) from west Africa, which was said to produce much better and more permanent reds.

A number of species of *Caesalpinia* have been recorded from Brazil, Colombia, Central America and the West Indies which have been called brazil since they all produce a similar red. The species that seems to have been most widely exported is *C.echinata*, which is a large tree with a red, thorny bark. It is also known as fernambuc or pernambuco, bois de brésil, nicaragua wood and St Martha's wood. *Caesalpinia brasiliensis* is also known as brazil wood and produces a red dye, but it is a small wiry shrub, so was probably not used commercially. *Caesalpinia crista* (illustrated here) is also known as brazil wood and also referred to in the literature as bois de pernambouc or fernambuco. It is a woody climber, with very spiny stems. A further identification problem arises with the dye wood known as 'brésillet' or 'brasiletto', which was imported into Europe and the United States from the

West Indies. Most authorities think that it was probably *Haematoxylon brasiletto*, but *C.bicolor*, *C.bahamensis* and *Peltophorum brasilense* have also been suggested. Pomet, writing in Paris in 1694, listed five kinds of brazil wood and recommends trusting to the honesty of the merchant for the best results. Pernambuco was said to have been the most satisfactory. Wood imported today for craft dyers is said to be *C.echinata* but we have not been able to confirm this.

Brazil wood is sold in the form of heartwood chips or occasionally as powder. The usual method is to pound the chips, soak them in water for one to seven days, and then use as for any other bark. A huge range of reds, pinks and purples can be produced, depending on the soaking time, the pH of the dyebath and the origin of the material. In our experiments, a startling range of colours was produced when small chips were treated with boiling water and then used immediately, after the scum had been removed with waste wool. In an acid bath, samples became yellow, bright orange, brownish-pink and dark tan with no mordant, alum, chrome and copper respectively. The addition of alkali gave deep crimson with alum, purple with chrome and near black with copper. Clearer bright colours are produced by extracting the pigments with alcohol. Cardon (1990) advocates boiling the dye for two to three hours, then allowing it to ferment for some days, while Bancroft (1813) suggests several months fermentation to produce bright reds. Other dyers state that adding gelatine or milk to the dyebath causes precipitation of some of the less desirable substances, resulting in much brighter reds. Sometimes a small piece of coarse cloth is dyed before the main batch to remove some of the impurities. The colours are not particularly light-fast. One pigment present in the wood is *brasilin* which is oxidized to form the red *brasilein*. Other pigments recorded are *haematoxylin*; *quercetagetin* and other flavonoids such as *fisetin*. *Haematoxylin* and *brasilein* were thought to be neo-flavonoids, but are now known to be more closely related to the true flavonoids.

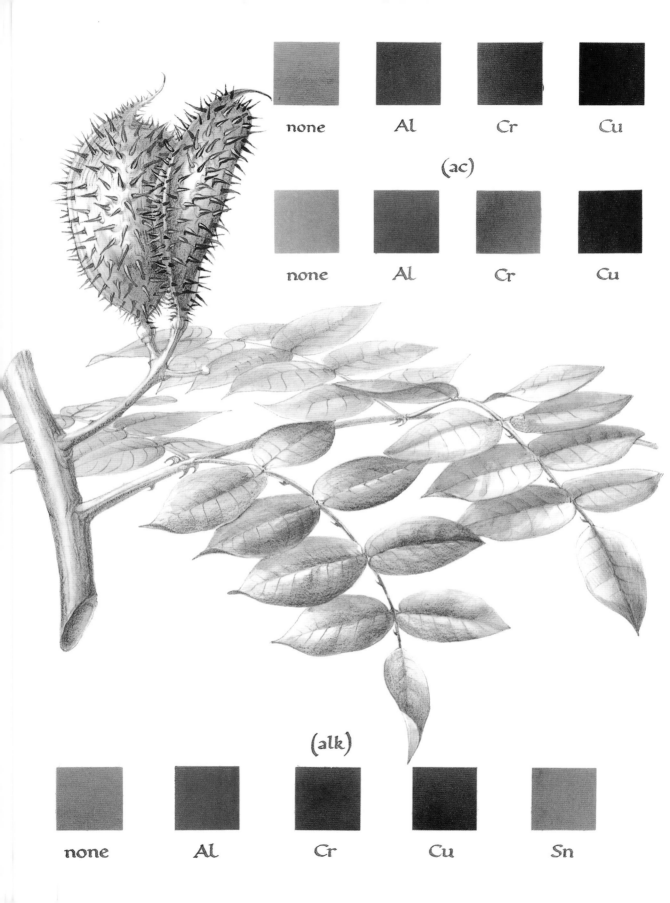

none   Al   Cr   Cu

(ac)

none   Al   Cr   Cu

(alk)

none   Al   Cr   Cu   Sn

Broom

Cytisus scoparius (Pea family – Leguminosae)

Broom is widespread throughout western, central and southern Europe. It forms a much-branched shrub up to 2.5m (8ft) high, with green, glabrous, five-angled twigs, lacking spines. The leaves are trifoliate, the leaflets elliptical to obovate, slightly hairy when young. The flowers are golden yellow, about 2cm (¾in.) across and the seed pods are hairy and black when mature. It grows on sandy, acid soils.

There are very few historic references in the dyeing literature to broom, perhaps due in part to the confusion over names (see below). It is sometimes recommended to be used together with saffron and elder, to give a golden tone to medium and light-brown hair, and to brighten faded blonde hair. An old recipe advises equal parts of elder bark, broom flowers and saffron, mixed with beaten egg yolks that have been boiled in water. This is applied to the hair but is said to be 'rather messy'.

Brooms are fast-growing and can be attractive in a small garden, although they have a short life-span. They will thrive anywhere except on chalky or limestone soils, growing best in a sunny position, where they can tolerate windy situations. They are best planted in autumn, but container-grown plants can be established at any time of the year, providing that the roots have plenty of water. Flowering takes place in May and June. The cultivar 'Andreanus' has attractive red and yellow flowers and produces good dye pigments.

The parts used for dyeing are the young stems before flowering starts and the flowers or flowering tops. Flowers are picked with as little stalk as possible and simmered together with the wool. A greenish-yellow is produced with alum, that fades to a yellower tint, and olive tones with chrome and copper, which fade a little in strong sunlight. The young branches are gathered in April or early May, chopped and used in the usual manner; greenish-yellows are produced with the usual mordants. The pigments in the flowers are flavonoids, of which the principal one is *scoparoside*, which is also present in the branches as is the isoflavone *genistoside*.

There is much confusion in dye literature between broom and dyer's greenweed (*Genista tinctoria*), which is sometimes given the vernacular name dyer's broom. In France particularly, the use of the common names causes difficulties because several quite distinct genera are all called 'gênet'. *Cytisus scoparius* is known as gênet à balai, *Genista tinctoria* is gênet des teinturiers and *Spartium junceum* is called gênet d'Espagne. The latter is also known in English as Spanish broom. All this provides an excellent example of the difficulties and confusion that can arise from the use of common names, without the back-up of the internationally understood scientific nomenclature.

Al

Cr

Cu

Coreopsis

Coreopsis tinctoria (Daisy family – Compositae)

Coreopsis, or tickseed as it is known in the United States, is a genus of about seventy species, found mostly in North America, although a few are native to Africa. *Coreopsis tinctoria* is widespread in the United States and Canada, usually occurring in open sandy situations, and in artificial habitats, such as roadsides and railway tracks. It is an annual with leaves which are in pairs, finely divided into very narrow segments. The flower heads are borne on long stalks, the disk florets a dark reddish-purple, while the ray florets are yellow or yellow with a red or brownish base.

Some plants in the same sub-group of this very large family have very similar dye properties. These are *Cosmos sulphureus* and *Dahlia pinnata*, both native to Mexico, and there is good evidence that Aztecs used them before the Spanish invaded Central America. Both species are grown in gardens, the latter is very familiar as the garden dahlia, which has a multitude of cultivars.

Coreopsis tinctoria looks elegant in the garden and is a commonly cultivated annual. It is easily raised from seed, sown in late April where it is to flower, or into seed pans with a normal compost under gentle heat. After germination the seedlings are then pricked out into seed trays and grown on in the usual way. The young plants thrive in any ordinary garden soil and should be thinned or planted out about 15cm (6ins) apart. In mild winters, a few plants may survive to the following summer, but do not rely on this. Many perennial species of coreopsis can also be used for dyeing, some of which can be bought at garden centres. Large, well-established plants can be divided and replanted in autumn. It is possible to propagate the perennial species by cuttings from young growth raised in a cold frame.

To encourage prolonged flowering, pick the flower heads each day; individuals left to seed do not flower as profusely. The flowers give very satisfactory results after drying for a few days in a warm airing cupboard, but make sure they are thoroughly dry before storing, as otherwise moulds may develop on the damp blooms. If any liquid containing pigments is left over after dyeing, this can be successfully frozen and used later. Small amounts of fibre can be dyed by simmering with the flowers; larger amounts are better if the flowers are removed before the fibre is added. Flowers grown during a hot summer produce excellent reds and oranges. The colours are altered by changes in pH, the best reds being produced in a slightly alkaline solution. The flowers should be simmered for a short time, until the maximum amount of dye appears to have been extracted. Dye the wool in the usual way and then lift out; add just enough washing soda dissolved in water to the dyebath to turn the liquid from yellow to red. The wool should be returned to the bath and simmered for a few more minutes until a good red is produced. Finally, rinse the wool in water with a little soda to maintain alkalinity. Some dyers who separate their material into flowers with maroon, orange or yellow ray florets, note that the colours produced vary, but we cannot confirm this. *Coreopsis tinctoria* contains several chalcones and aurones, called *coreopsidoside* and *sulphuroside*, *mareoside* and *maritimetoside*, and also anthocyanins.

Cu

Cr

Sn

Cr+Fe

Al

Cutch

Acacia catechu (Pea family – Leguminosae)

Acacia is a large genus of over 500 species of mostly evergreen trees and shrubs. They occur in many of the warmer regions of the world, although most are found in Australia. *Acacia catechu* is a deciduous tree of up to 25m (80ft) in height and is native to India, Burma and Sri Lanka, and cultivated in many tropical regions. The tree is spiny with downy young shoots and dark bark which hangs peeling in long strips. The compound leaves bear up to forty pairs of short, narrow leaflets. The numerous flowers are small, dark yellow, and are arranged in spikes of 5 to 12.5cm (2 to 5ins) in the leaf axils. The fruit pods are long and strap-shaped.

Cutch was first used in Europe in the 1800s, but had been frequently used in India long before this, for dyeing cotton. It was being exported to China and Japan in the early 1500s, according to Barbosa, a Portuguese explorer, and was sometimes referred to as 'terra japonica', because it was thought to be an earth found in Japan. In 1574, Garcia de Orta wrote a remarkably complete account of cutch production and use in India (Watt 1889–93). A number of other species were used in the same manner as cutch and all were referred to as catechus. *Acacia catechu* was sometimes known as 'bengal catechu'; *Areca catechu*, the betel nut from the palm family, was called 'bombay catechu' and *Uncaria gambier* of the madder family, called 'gambier catechu' or 'gambier', was the source of gambier extract. The dye of *Areca catechu* is prepared from the palm nuts and is still used in the Philippines, together with flowers of *Toona ciliata* from the mahogany family, or roots of *Morinda citrifolia* of the madder family, to give red dyes, or black dyes with an iron mordant. The betel nuts are chewed by many people throughout Southeast Asia. Gambier is prepared in a similar manner to cutch, using the leaves and long, trailing branches. It is sometimes referred to as 'pale catechu', and contains large amounts of catechols and tannins.

Cutch is supplied in the form of a coarse powder. It is prepared from the chipped heartwood of the trees, which is macerated in water by boiling and reboiling the resulting mash. When the right consistency has been reached, the syrupy liquid is poured into moulds and sets into a dark, caramel-smelling block. This is subsequently pounded to form a coarse powder which is sometimes termed cutch-extract. On wool, a light tan can be produced without a mordant, and a strong tan, deep reddish-tan and chestnut brown with alum, chrome and copper respectively.

Cutch is an excellent dye for wool, silk and cotton. For wool, dissolve cutch equal to 5 per cent of the fibre weight in very hot water. Add this to the dye bath, together with enough cold water to allow the wool to move freely, with or without the addition of copper sulphate as a mordant (6 per cent of the weight of the cutch used). The fibres are dyed in the usual way, keeping the water temperature below 80°C (176°F). Do not allow the fibres to cool in the dye bath. The colour is fixed by oxidation which occurs during rinsing, or during a further mordanting with chrome (1 to 2g (1/16oz) per litre of water). Cutch is an extremely permanent dye when used in the way described, with an excellent colour- and wash-fastness.

The dye principles involved are flavonoids including *quercetagetin*, with large quantities of catechols and tannins.

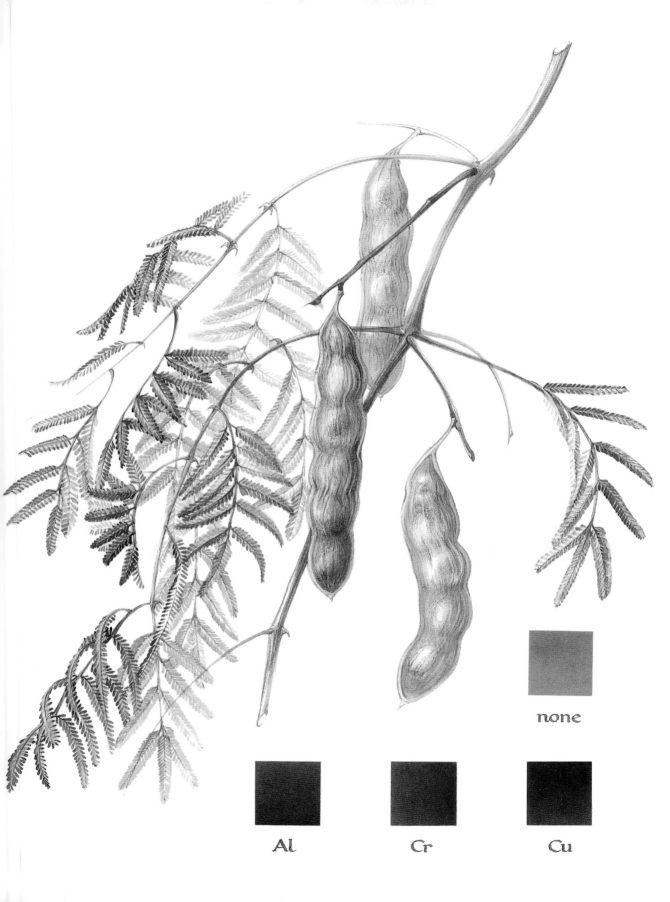

Al

Cr

Cu

Dyer's alkanet

Alkanna tinctoria (Borage family – Boraginaceae)

There are about twenty-five species of *Alkanna*, all native to southern Europe and the Middle East. Dyer's alkanet is a low-growing, bristly perennial, with stems prostrate or ascending. The linear-lanceolate basal leaves are 6 to 15cm (2½ to 6ins) long and up to 1.5cm (⅝in.) wide. The flowers are a vivid blue, 6 to 7mm (¼in.) across and the fruit are four small, greyish nutlets, located within the calyx. It thrives in hot sandy, rocky places, and is sometimes abundant at the rear of sandy beaches, where it forms low domes up to half a metre (1½ft) across. In spring, they are covered with gentian-blue flowers which make a spectacular splash of colour. The parts used for dyeing are the roots. Do not confuse this plant with the *'Anchusa'* of gardens (*Pentaglottis sempervirens*), which is often used by mistake for alkanet with disappointing results.

Alkanet was the principal adulterant used with the true purple dyes from shellfish and its use is mentioned in Greco-Egyptian papyri from the third century AD. It was used for cosmetic purposes in ancient Greece and Rome, and as a wine colorant until very recently, appearing in the provisional list of permitted colorants by the EEC, but subsequently disallowed. It was sometimes used to colour the fancy display bottles in chemists' windows, and in thermometers. Bancroft (1813) records its use primarily as a source for a beautiful purple-violet on silk. He also refers to *Anchusa virginica* (Puccoon) which possessed a similar colouring matter, and was formerly used by American Indians to paint their bodies. The name *Alkanna* is derived from the Arabic *al henneh* and the similarity of this name to *henna* has given rise to a lot of confusion in the literature. The *'Alkanna'* mentioned by Adrosko (1971) from the southeast coast of the United States must be a species of another genus, as *Alkanna* is confined to the Old World. The *Anchusa* mentioned by Bancroft is now known as *Onosmodium virginicum*. True *Alkanna* has so much pigment in its roots that the heavy paper used to mount museum specimens is often heavily stained with purple; there is no staining visible on mounted sheets of *Onosmodium* in The Natural History Museum, London. A related genus, *Lithospermum* is well known in both the Old and New Worlds for the red dye in its roots, and two species from the United States are correctly called puccoon. The museum has a number of badly stained sheets; one of *L.incisum* bears the information 'Indian paint or puccoon', another notes a royal purple dye from the roots. *Lithospermum caroliniense* and *L.arvense* sheets also show colour staining. Bancroft probably relied on the names being given to him for material sent from the United States and it is likely that the confusion arose in this way. We think it is highly probable that both Adrosko's *'Alkanna'* and Bancroft's *'Anchusa'* were both a *Lithospermum* species.

Alkanet is usually sold as chopped roots. It is rarely available as plants for the garden. If you are fortunate enough to obtain some, it prefers a very light sandy soil free from lime, in a warm position. However, it may not survive a very cold winter. The principle dye substance – *alkannin* – is not soluble in water, but can be dissolved in alcohol, acetone and various fats. The easiest method is to use colourless rubbing alcohol from a pharmacy or extraction with methylated spirits. A very small amount of the latter extracts a lot of the dye. The synthetic purple dye in methylated spirits is very dilute and can be safely used without altering the colour. Test skeins in dilute methylated spirits showed virtually no colour change. About 50g (1¾oz) of alkanet roots will dye 100g (3½oz) wool to quite deep shades when used with this extraction method and various greys, violets and purples can be obtained. Alum gives a beautiful violet-purple, chrome a violet-slate and copper a shade of brown. Unmordanted wool is dyed a pretty blue-slate which totally disappears in alkaline or acid rinses. None of the colours are very permanent. The pigments present are napthoquinones, of which *alkannin* is the most abundant.

Al

Cr

(alk)

Al   Cr   Cu

Dyer's chamomile

Anthemis tinctoria (Daisy family – Compositae)

Dyer's chamomile grows in dry places and is found over much of Europe, particularly the south, extending eastwards to the Himalayas. It sometimes occurs, as an escape from cultivation, in the British Isles and in North America. This species was not much used in Europe, as weld and saw-wort were both thought to be much better sources of yellow. It was used in Turkey for carpet production before synthetic dyes became popular.

Stinking mayweed (*Anthemis cotula*) and corn chamomile (*A. arvensis*) both have flower heads with white rays and yellow centres and can be used in a similar manner to dyer's chamomile. Related genera contain species with similar dyes, though none of them give quite such strong colours. These include the common chamomile (*Chamaemelum nobile*) which can be used fresh or dry. It can be bought rather expensively, as chamomile tea, from supermarkets and health-food shops. Pineapple weed (*Matricaria discoidea*) and scentless mayweed (*Tripleurospermum inodorum*), often found on track sides and waste places, give similar results.

Dyer's chamomile, a perennial, grows up to 60cm (2ft), with much divided leaves that are often white-hairy beneath. The flower heads are 2.5 to 4cm (1 to 1½ins) across, borne singly on long stalks. The receptacle is hemispherical, with lanceolate scales amongst the florets. The petal-like ray florets are yellow and all female, the hermaphrodite disk florets are also yellow. The achenes are four-angled, hairless, and have a very short border or rim at the top. The only plant that dyers in the British Isles are likely to confuse with this species is corn marigold (*Chrysanthemum segetum*). This was formerly a widespread arable weed, but modern agricultural methods have now made it uncommon. The similarity lies with the flowers, which have yellow ray and disk florets, but a glance at the leaves will distinguish the two species. Corn marigold has leaves that are completely glabrous and a rather bluish-green, only coarsely toothed and the upper leaves are stem clasping; the receptacle is convex, without scales.

Dyer's chamomile is a frequently planted border plant and can be bought from many garden centres; several varieties are available. It is easily propagated from seed or by dividing rootstocks in autumn or spring. It thrives in any well-cultivated soil but is somewhat intolerant of lime, and needs plenty of sun. Plants grown in the shade seem to produce less dye than those from sunny positions. Flower heads should be picked regularly and can be dried in the airing cupboard in the usual way; they are also available from suppliers of dyestuffs. The leafy green parts of the plant also contain much dye material and should be picked when the stems are mature.

Alum-mordanted wool becomes a beautiful, soft golden-yellow; chrome gives a rather duller yellow; and copper a greenish-yellow. Unmordanted wool is scarcely coloured at all. The colours fade a little in strong sunlight, wool mordanted with alum fading the most. The pigments present include the flavones *apigenin* and *luteolin* and the flavonols *quercetagetin* and *patuletin*.

Al

Cr

Cu

Dyer's greenweed

Genista tinctoria (Pea family – Leguminosae)

Dyer's greenweed, or woad-waxen, is native to most of Europe, the Caucasus and Asia Minor. It is a very variable, small shrub, rarely with small spines, and has spreading stems that may reach 2m (6ft), but is usually much lower. The oblong-lanceolate leaves are up to 5cm (2ins) long and glabrous to densely silky-hairy. The pale yellow flowers are shortly stalked and the petals do not fall until the pods are mature. In the British Isles it flowers in July and August.

Dyer's greenweed was used chiefly for giving a yellow base which was then top dyed with woad or indigo to give green. The vernacular names of dyer's greenweed and woad-waxen reflect this use. It was sometimes used with the addition of weld, together with woad, to give a colour known as Kendal green. Lincoln green seems to have been produced in medieval times by first dyeing the cloth blue with woad, then over-dyeing with dyer's greenweed or weld. Textile historians believe that the green wools in the Bayeux Tapestry were produced with this dye in conjunction with woad. As the tapestry is generally considered to be of English origin, it must be one of the earliest examples of the dyer's craft from Britain. The Swedish botanist Linnaeus mentioned its use as a dye plant in 1749. Bancroft (1813) refers to the production of 'an inferior yellow' on coarse woollens that had been mordanted with alum and cream of tartar. Other species of *Genista* which can be used for dyeing are the hairy greenweed (*G.pilosa*), a plant of sandy heaths and poor soils, which gives an acid yellow to wool mordanted with alum; the petty whin (*G.angelica*); and the related *Chamaespartium saggitale*, which is often grown as a rock garden plant.

All give yellows of various intensities.

Dyer's greenweed can be easily bought from garden centres and several, usually non-spiny, varieties are available. It is also sold under the names of *G.depressa* and *G.humilis*, which are regarded as synonyms by botanists. In the wild, the plants grow in rough pastures and along roadsides, often on clay soils. Any kind of garden soil suits these plants, but like many other dye plants, more pigments appear to be produced in the sun than the shade. Plants can be grown from seed, which sould be lightly filed before sowing. This breaks the hard seed coat and increases the chances of germination. Seedlings should be planted out where they are to grow to maturity, and spaced about 30cm (1ft) apart.

The dye is concentrated in the flowers, which should be picked at maturity. They are best used fresh or can be used after drying. The flowering tops can be picked in June or July. To get a strong colour with dyer's greenweed, use about three to four times the weight of flowers to wool. For the best colours, exclude the leaves and stems from the dye bath, although these can be added if sufficient flowers are not available. Unmordanted wool becomes a faint yellow which is fairly light-sensitive. The strongest acid-yellow is produced on wool mordanted with alum. Rather surprisingly, wool mordanted with chrome, using flowering tops with few leaves, becomes a strong orange, while copper gives a dirty greenish-yellow. The colourless pigments present in the plant form coloured derivatives in the dye bath which are the flavone *luteolin* and an isoflavone, *genistein*.

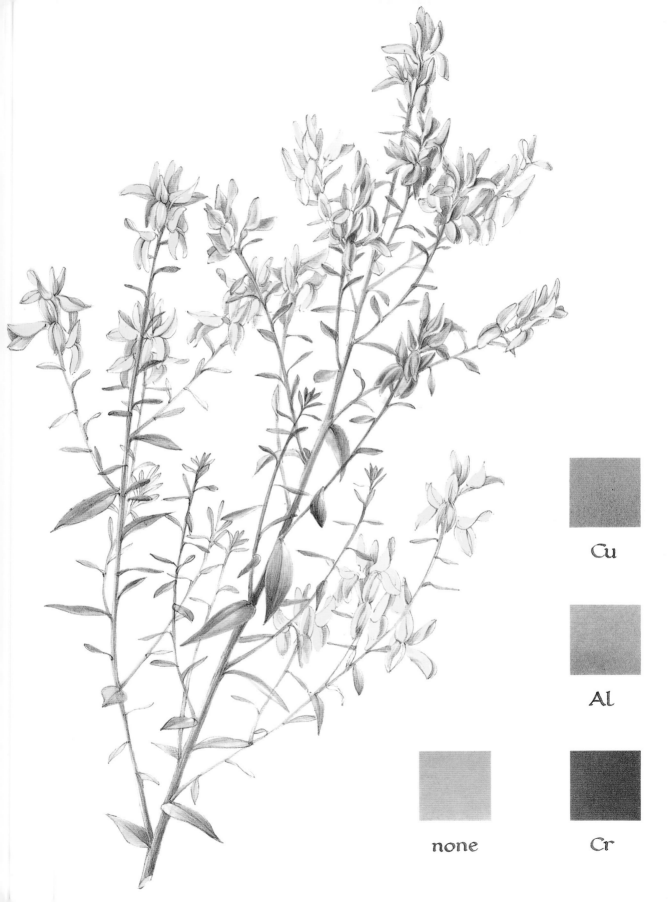

Cu

Al

Cr

Elderberry

Sambucus nigra (Honeysuckle family – Caprifoliaceae)

Elderberries are native to most of Europe (except the extreme north), western Asia and North Africa. They are common in wood margins, hedges and waste places, particularly on disturbed base- and nitrogen-rich soils, such as those round badger sets and rabbit warrens in southern England. Elderberries are shrubs or small trees of up to 10m (33ft), with branches that are often arching and have a pronounced central pith. The leaves have 5 to 7 slightly toothed leaflets, and the numerous, small, creamy-white flowers with yellow anthers are heavily grape-scented, and are borne in flat-topped clusters. The berries are black and juicy and, in the British Isles, mature during August and September.

The dye from the berries was used with copper in Germany in the early eighteenth century to give a blue to linen. It has also been used to give a mauve colour to cotton and is sometimes used to improve the colour of grape wine, in addition to its familiar role in elderberry wine. A modern recipe uses the berries to give silver-blue or violet tones on yellowish hair, although it is wise to omit the leaves because they may give the hair a greenish tinge.

There are several varieties of elderberry, many of which can be bought from garden centres. Some varieties have deeply toothed leaves (*var.laciniata*) and others have variegated or golden leaves. All thrive in any good garden soil. Leaves and berries are widely used and dyers occasionally use the bark and roots. Materials can be gathered from hedgerows, the leaves at almost any time before they fall, and berries should be fully ripe, but not frosted. Berries are best used fresh, but can be dried in a cool oven for later use. Storage by freezing is not satisfactory. Colours from leaves vary with the season; in mid-summer, shades given to wool are very pale greens with alum, gold with chrome and olive with copper; but in late autumn a soft green is produced on copper-mordanted wool. The ripe berries give pinks, violets and blue-greys, depending on the mordants. With alum the colour is a light grey, bluish with chrome and a slatey-grey with copper. Adding vinegar shifts the colours towards violet-reds, while salt shifts the colours towards bluer ranges. Using vinegar alone with unmordanted wool gives a pinkish-purple. The colours are not very light-fast and copper as the mordant gives the best results. To make the colour more permanent allow the wool to cool in the dye bath and remain in the liquor overnight. The pigments contained in the berries are the anthocyanins *chrysanthemin* and *sambucin* which are derivatives of *quercetagetin*. Tannins are also present.

Dane's elder (*Sambucus ebulus*), is a shorter hedgerow shrub, with leaves having 7 to 11 leaflets. The flowers have purple anthers and an unpleasant smell. It has been suggested that the berries might have been the source of the blue pigment used by ancient Britons to dye their skins; but because the dye is grey-pinkish with acid on unmordanted wool, the natural acids on the skin would presumably give similar results, and it seems more likely that woad, usually noted in this context, was used. The juice is used illicitly to colour wine despite its strongly purgative qualities.

The ripe berries give various pale pinkish or greenish-greys with the usual mordants in an acid bath; all are very light-sensitive except when iron is used as the mordant. In an alkaline bath the colours are much deeper, shifting to greens, which again are very light-sensitive, except for those produced with copper which remain quite stable. The wool fades quite badly when iron is used.

The leaves and branches of the red-berried elder (*S.racemosa*), which is sometimes naturalized, particularly in Scotland, have also been used to give a golden brown with alum, that is delicately known in France as *merde d'oie dorée*.

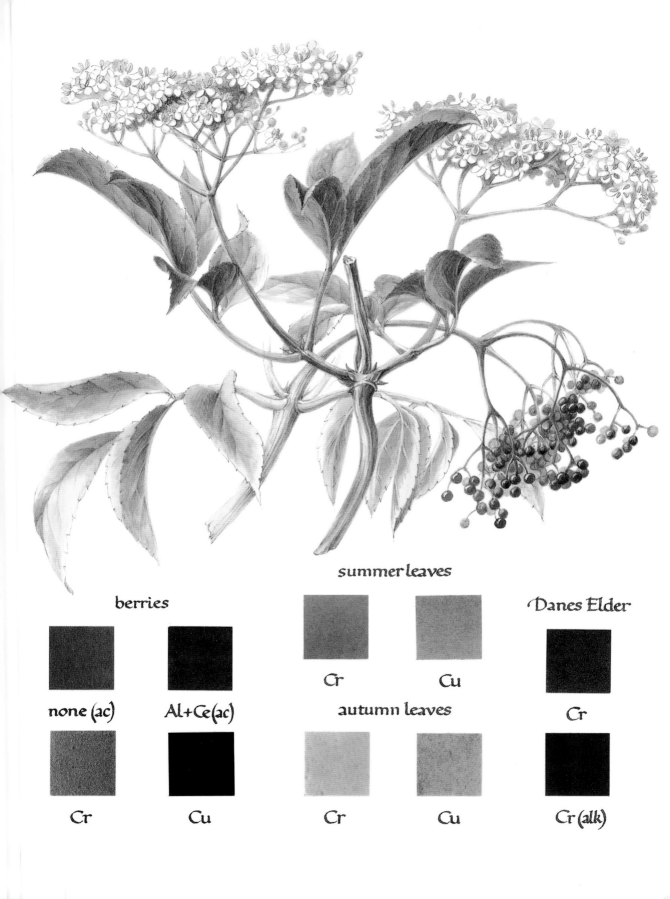

summer leaves

berries

Danes Elder

Cr   Cu

none (ac)   Al+Ce (ac)

autumn leaves

Cr

Cr   Cu   Cr   Cu   Cr (alk)

Eucalyptus

Eucalyptus gunnii (Myrtle family – Myrtaceae)

*E*ucalyptus gunnii is native to the mountains of Tasmania. It grows up to 25m (80ft), and has a bluish foliage. Juvenile leaves are round and silvery blue, while those on mature growth are long, sickle-shaped and somewhat greener.

This species can be bought as small trees or bushes from garden centres, but it is not difficult to grow from seed. Keep the seeds in a warm place after sowing until germination. The trees thrive in most soils and are moderately hardy, but can be severely damaged by hard and prolonged frosts. In areas where many frosts occur, or where it is inappropriate to plant a tree, it can be kept as a shrub by regular pruning. This also preserves the juvenile foliage which is particularly valued by flower arrangers. Both trees and shrubs are very fast growing.

Leaves and young branches can be used fresh or dried – a few days in the airing cupboard not only preserves the leaves, but also gives a pleasant fragrance to the airing clothes. Use 200g (7oz) of dried leaves to 100g (3½oz) wool for strong colours. Bark and fruit give more muted tones. The colours produced vary considerably according to the seasons. In mid-summer, leaves give soft oranges and browns; unmordanted wool and alum give very similar oranges with pinkish tones, chrome a pale yellow-orange, and copper pinkish-brown. In late autumn, much of the red in the dye disappears and the colours are much more in the yellow range. The pigments are flavonols and are very light-fast. The flavone *hemiphloin*, rhamnosides and tannins are also present.

Many other species of *Eucalyptus* can be used for dyeing, including *E.cinerea*, *E.crenulata*, *E.nicholii*, *E.globulus*, *E.bicostata* and *E.cordata*, most of which are only doubtfully hardy in Britain. *Eucalyptus cinerea* and *E.cordata* both give red colours, the former is sometimes used in textile painting and printing. *Eucalyptus cordata* is known as the silver-dollar eucalyptus and is sometimes sold by florists for flower arrangements. *Eucalyptus globulus* is frequently planted in southern Europe for paper pulp; it is also used as an ornamental bedding plant, particularly in large displays, because the juvenile shoots are an attractive blue colour. It can easily be grown from seed, but should not be planted out until all danger of frost is past. It will sometimes survive mild winters in the south of England and very quickly grows into a small tree. Minor frost damage will kill the leading shoots, but young growths may then shoot out of the damaged branches. Green and grey colours can be obtained from this species. The bark of *E.macro-rhyncha* is much used in Australia for its tannins; its leaves contain *rutin* which is a derivative of the flavonol *quercetin*. According to recent experiments in Australia, Cockburn & Jost in Buchanan (1990), all 600 species of eucalypts give colour from their leaves, with a range from pale yellow to dark red. Barks and fruit usually give reds, greys and browns, although the colours vary markedly with the season, temperature and rainfall. Details are given of the methods used. For example, dark reds were produced from a number of species, by boiling the leaves in the dyebath for several hours, then adding unmordanted wool. *Eucalyptus globulus* gave green with alum and chrome, as did several other species. References are given in this article to further literature concerning dyeing with *Eucalyptus*.

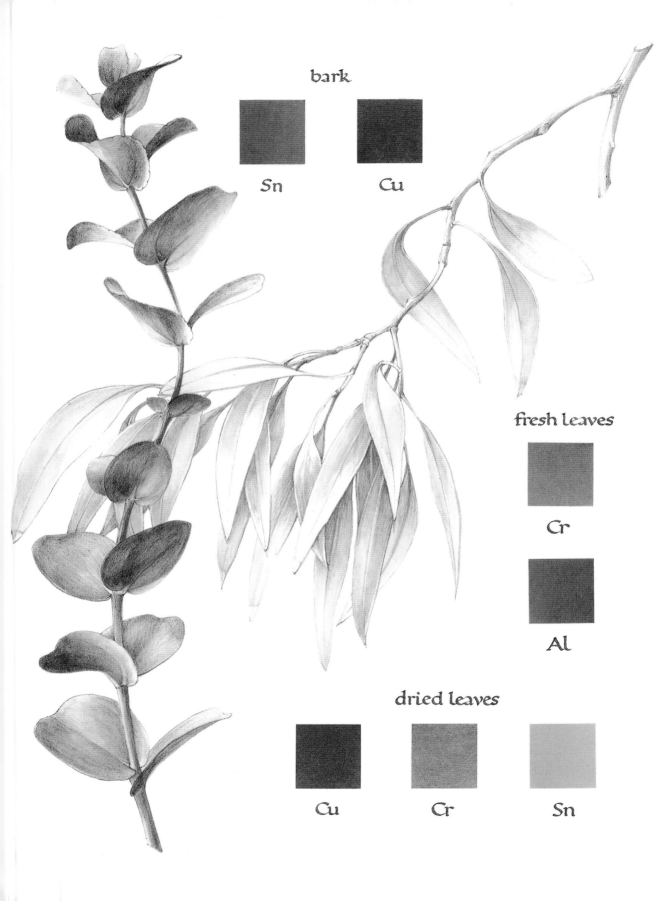

bark

Sn   Cu

fresh leaves

Cr

Al

dried leaves

Cu   Cr   Sn

French marigold

Tagetes patula (Daisy family – Compositae)

French marigolds, despite their common name, are native to Mexico. Most other species of *Tagetes* are found in central and south America and several of them are also called French marigolds. The commonly grown species are annuals and grow up to about 45cm (18ins) high. They have much-divided leaves with toothed segments and the flower heads can be yellow, orange, red or brownish and are borne singly on long stalks.

French marigolds are extremely popular garden annuals and are planted for decorative purposes through the temperate regions of the world. They are very easy to grow from seed and should be germinated in a warm place. The seedlings should be pricked out into boxes and grown on in a cold greenhouse, but they should not be planted out until all danger of late frosts is past. They are also widely available ready grown from garden centres. The plants will continue to flower for a long period if the old blooms are removed regularly.

Mature and somewhat withered flowers are used for dyeing, and even frosted blooms give excellent colours. They may be used fresh or dried in the usual way. Unmordanted wool gives a pale yellow, alum a golden-yellow. Chrome gives dark orange and copper gives brownish tones. Different times of the year and possibly different horticultural varieties give various yellows, oranges and tans, or even olive-greens. The colour-fastness is good, but colours may darken a little, particularly the olive-greens, which become brownish. The pigments present include two derivatives of *quercetin* – *quercetagetin* and *patuletin*.

The African marigold, *Tagetes erecta* is larger than the French marigold and grows up to 50cm (20ins). The flower heads are also bigger, being yellow to orange and without the red or brown tints of French marigolds. The colours produced include a brilliant orange on wool or silk mordanted with tin, which is very light-fast. The pigments present are *quercetagetin* as in the previous species and the carotinoid *lutein* (known to the European food industry as E.161(b)). Hybrids between the two species are available from most nurseries, but their dye properties are not known to us. *Tagetes micrantha*, known in the United States as bitterball, is a native of the south-western states, and has very small flowers, often with only one ray floret. Despite its small size, it apparently can produce good colours on wool, with the dye pigments probably concentrated in the leaves. It is recorded as giving a bright yellow-green with alum and brassy yellow-green with chrome after simmering the flower heads in the dye bath at 80°C (176°F) for two hours prior to dyeing.

Cr

Al

Cu

Fustic (Old fustic)

Chlorophora tinctoria (Mulberry family – Moraceae)

Fustic is a native of tropical America. It is a spiny tree up to 30m high with wide-spreading horizontal branches. The leaves are lanceolate to elliptic, 5 to 12cm (2 to 4¾ins) long, on short stalks. Male flowers are borne in pendulous pale greenish-yellow catkins up to 10cm (4ins) long; the female flowers, which are green, are arranged in globose heads of up to 10mm (⅜in.) across, on short stalks. The fruit is up to 14mm (⁹⁄₁₆in.) in diameter, fleshy, grey-green in colour and edible.

Fustic is cultivated in the West Indies, Central and South America and historically the wood imported from Cuba was considered to be superior to the rest. Another member of the genus, *C.excelsa*, known in commerce as 'iroko', is imported from Africa as a teak substitute. Fustic dyestuff was first introduced into Europe via Spain in 1510 and was later imported directly by several countries in very large quantities. In Britain, it became very important when khaki replaced the bright colours of army uniforms. It is one of the fast-growing trees that can be used for reafforestation, and has recently been planted in Colombia in areas devastated by natural or man-made disasters. The wood is still in demand and could become a truly renewable resource.

Fustic is imported as logs with the bark removed. It is a bright yellow wood and is used in the form of chips or after it has been pulverized. The chips should be soaked in water for several days before use, because the dye runs more easily from the softened wood. The dye is substantive, though unmordanted wool becomes a paler yellow than that mordanted with alum. Chrome gives a yellowish-orange and copper a brown-tan. Orange-tans and browns can be produced by simmering the wool for longer periods than is needed to obtain the yellows. Bancroft (1813) says that if the tannins present are precipitated (with gelatine or glue) the yellows produced will be much brighter. The colours are very light-fast, but there may be a little darkening if they are exposed to very strong sunlight. Fustic can be used in the production of other colours, for example, with indigo or woad for greens. It is sometimes used to give a base for a black dye with logwood, using a chrome mordant for wool and iron for silk, The pigment present is the flavonol *morin*, together with a large quantity of tannins. Where fustic is still used commercially for yellows, the tannins are removed by precipitating them with gelatine before dyeing.

The white mulberry, *Morus alba*, is a relative of fustic. It is a native of central and eastern China and is widely planted in Europe and North America. It grows into a small tree and flourishes in chalky or limy soil, as well as on gravels and sands. Its leaves are used in sericulture as the preferred diet of the caterpillars of the silk moth. Different races of silk worms are able to absorb varying quantities of the flavonoids present in the leaves, so that the resulting silk can vary from white to a moderate yellow. The main pigment present is the flavonol *morin* and *rutin* is also present. The bark and wood contain various flavones of which *mulberrin* is one; strong golden yellows, bronze, olive and browns can be produced with varying mordants. The fruit of the white mulberry are edible, but those of the black mulberry are more highly esteemed for culinary purposes.

Al

Cr

Cu

Goat willow

Salix caprea (Willow family – Salicaceae)

The goat willow is native to Europe and is also found through Turkey into Central Asia, with very close allies in China and Japan. It is locally common in the British Isles, where it is found in hedgerows, wood margins and on rocky lake shores, and tolerates dryer situations than many other willows. It was a common invader of bomb sites during and after the Second World War, and is often found along railway banks. Goat willow is often gathered as 'palm' and used in decorating churches for Palm Sunday.

Willow bark has been used for many centuries, particularly for tanning, and is referred to in Egyptian papyri from 2000 years ago. It is still used today to produce powdered tannin extracts, as well as for dyeing and medicinal purposes. Other species of willows recorded as being used for dyeing include *S.viminalis* – the osier, *S.purpurea*, *S.alba* and *S.fragilis*, the latter three also being the source of salicin, from which the familiar household drug aspirin was derived before it was manufactured synthetically.

Goat willow is one of the earliest willow species to come into flower, the familiar catkins appearing before the leaves. It forms a bush, or tree of up to 10m (33ft). The leaves are shortly stalked, with little ear-shaped projections at the base of the stalks, which are shed before the ovate or often almost orbicular leaves are fully developed. The male and female flowers are found on different bushes. The male catkins are up to 2.5cm (1in.), white and silky at first, later becoming golden-yellow as the numerous stamens mature and elongate. The female catkins are also silvery, about the same size as the males, but elongating and greenish in fruit. Goat willow sometimes hybridizes with other species and although we have not investigated the dye properties of the hybrids, they are unlikely to be very different. Goat willow can be distinguished from another early flowering species of rather wetter areas, the common sallow – *S.cinerea*, by peeling a twig. The wood of the goat willow is smooth while the sallow's is ridged.

Many species of willow can be bought at garden centres. The pendulous variety of *S.caprea*, which is known as 'Kilmarnock', becomes a small, umbrella-like tree, rarely reaching 3m (10ft), and is quite a striking feature for a small garden. The older twigs should be thinned out regularly, otherwise the graceful branches produce only an inelegant dome of greenery.

The bark of willows is most often used for dyeing, but leaves and young branches are also used. Bark is best soaked for several days before use. The innermost parts of the bark from a mature tree can produce excellent results after soaking for a week or two and bark from peeled twigs also produces good colours. Young bark from twigs peeled in midsummer gives tans, while unmordanted wool turns a rather pinkish-tan. Alum, chrome and copper produce less pinkish tans and iron gives various greys and grey-browns. Mature bark gathered in February gives a remarkable range of colours from golden tan to pinky-browns with alum, chrome and copper, while adding iron turns unmordanted wool a beautiful greyish-green. Alum and chrome with iron give attractive greys, and copper and iron give a very dark brown. Leaves give various shades from pinkish fawn on unmordanted wool, pale golden-brown with alum, a brassy colour with chrome and a brown-tan with copper. All the colours seem to be remarkably light-fast. The pigments present in the bark are chalcones, together with abundant tannins, while in the leaves there are flavonols and smaller amounts of tannins.

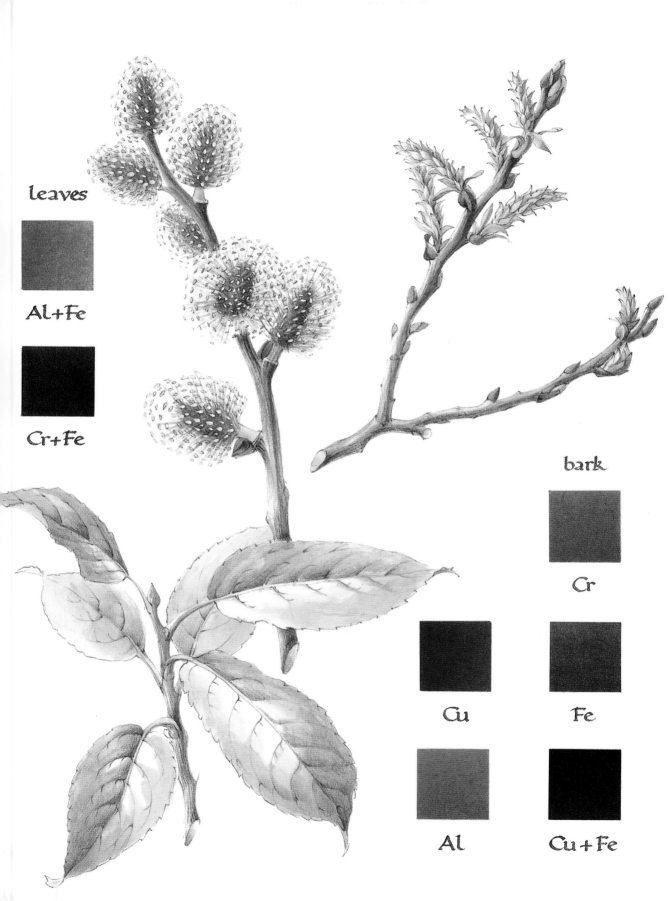

leaves

Al+Fe

Cr+Fe

bark

Cr

Cu

Fe

Al

Cu+Fe

Goldenrod

Solidago canadensis (Daisy family – Compositae)

There are about 120 species of goldenrods, nearly all of which are confined to North America. The common goldenrod of our gardens is *S.canadensis*, a perennial of about 1 to 1.5m or even up to 2.5m (3½ to 8ft). The leaves are lanceolate, three-veined and roughly hairy. The very numerous flower heads are golden yellow, up to 5mm (⅜in.) in diameter, and form pyramidal inflorescences. The petal-like ray florets are small and scarcely longer than the disk florets. Goldenrod often escapes from cultivation and may cover quite large areas of waste ground. Other similar species also occurring as garden-escapes include *S.gigantea*, a more robust plant with petaloid ray florets much longer than the disk florets, and *S.graminifolia* with smooth, narrow, five-veined leaves and ray florets not longer than the disk florets.

Solidago virgaurea is the one species of goldenrod native to Europe; it also occurs in northern and central Asia, and North America. It grows to about 75cm (30ins) high, but it is often much shorter. The leaves are ovate to lanceolate and shortly stalked. The individual flower heads are much larger than those of *S.canadensis*, with a diameter of up to 10mm (⅜in.) and the ray florets are much longer than the disk florets. There are far fewer heads in the much more loosely arranged inflorescences. This species is found in wood margins, on dry banks, on mountain ledges and usually occurs on acid soils. Two of the common names of *S.virgaurea* are Aaron's rod and Jew's rod, which refer to its use to dye the distinctive yellow clothing Jews were forced to wear during the Middle Ages. Goldenrods are thought to have been used by the Navaho Indians in North America before the arrival of Europeans. Bancroft (1813) tried to introduce the American plants for the calico printing industry, because he thought they would have been a good substitute for weld. He found they produced good yellows on silk and cotton when mordanted with alum, but they were not generally accepted by the dyeing industry.

Goldenrods flourish in almost any soil and may be very easily grown in the garden. Several varieties can be bought at garden centres. Those with smaller gardens might prefer *S.flexicaulis* or one of its many varieties, most of which are much smaller plants and almost certainly have similar dye properties to *S.canadensis*. Seed of the shorter varieties is sometimes available.

Goldenrods should be picked when the flower heads are in full bloom and can be used fresh. Dried heads can be used but give progressively less colour as they age. Using the flower heads alone produces a clearer yellow, but there are also dye materials in the stems and leaves. Adding alum to the dye bath produces a strong golden-yellow, which is more greenish when the leaves are present. Using chrome gives a brassy orange, while copper gives a yellowish buff, being rather more brownish when leaves are present. Tin produces a vivid pure orange, iron a khaki. These colours are reasonably permanent. *Solidago virgaurea* is used in exactly the same way as *S.canadensis* and gives similar colours.

The pigments in goldenrod species include the flavonols *kaempferol* and *quercetin*, the anthocyanin *cyanidin*, with tannins also present.

Sn

Al   Cr   Fe

Heather (Ling)

Calluna vulgaris (Heather family – Ericaceae)

Heather is a common plant of western Europe, becoming less common towards the south and rare in the Mediterranean region. It also occurs in North Africa, western Asia and the Azores and has become naturalized in a few parts of the eastern United States. It is a low shrub, rarely reaching 1m (3¼ft) high, and occurs on heaths, moors, bogs and in open woodlands on acid soils. The stems are wiry and root at the base; the leaves are 1 to 2mm (¹⁄₁₆in.) long, in opposite pairs, crowded on the shoots, often in four rows and sometimes hairy. The flowers are pinkish-purple (rarely white) and shortly stalked. Flowering is from July to September.

Heather was used in Scotland in the production of yellow wools for tartans in the eighteenth century, and was also used as the base for a green colour, being overdyed with indigo. Bancroft (1813) says it produces a rather fugitive yellow which can be made more permanent when tin is used as the mordant.

Heather can be gathered on moorland where it is abundant. It is sold in garden centres, and a large number of varieties are available, including some with double or strikingly coloured flowers, coloured leaves, and dwarf or tall habit. If you buy heather seed, it should be grown in special 'heather compost', but it is better to get cuttings from friends. These root very easily and grow quickly into large plants. Mature plants should be kept well pruned as they become very woody with age and produce fewer young shoots which are required for dyeing. Most heathers will not grow on limy or chalky soil and rooting composts must be lime free. In chalky districts use rain-water, rather than tap-water.

Gather the heather just before flowering. The young tops should be pulled or cut from the plants and the leaves and flower buds removed from the stalks by pulling them through the prongs of an old table fork. This material can be used fresh, or dried and stored for later use, as very little colour loss occurs during drying. It is also possible to use flowering spikes; some dyers even use old flower heads collected at the end of the year.

Unmordanted wool becomes a very attractive pinkish colour when dyed with pre-flowering tops and stems. Alum and tin give a golden-orange tone, chrome a tan and copper chestnut-brown. Unmordanted wool and the alum- and chrome-mordanted wools become somewhat paler in strong sunlight, while those mordanted with copper scarcely change. The addition of iron to the dye bath gives a grey-fawn on unmordanted wool and shades of greyish-brown with mordants. Wool dyed without the stems becomes yellow with alum, which is enhanced when tin is added, although this makes it more light-sensitive. The pigments present in the flowering tops include the flavonols, *quercetin* and *myricetin*; tannins are present in the woody parts of the plant.

Other species of heathers commonly found in the British Isles are bell heather (*Erica cinerea*) and the cross-leaved heath (*E.tetralix*). They have much larger flowers than the common heather and flower about a month earlier. Bell heather has showy spikes of reddish-purple flowers, and the cross-leaved heath has clusters of pale-pink flowers at the top of the flower stems. Both are found on moorland and give similar colours to common heather; they have been used in Scotland for tartans. Dyers with chalky gardens may like to experiment with *Erica herbacea* (often sold as *E.carnea*) which is available in numerous varieties and hybrids, and is the only readily available heather that will tolerate lime. A Mediterranean heather, recorded as *E.mediterranea*, but more probably *E.arborea*, was used in Greece to dye wool for much-admired rugs. The colours were produced by cold dyeing, involving fermentation of the boiled plants until 'thoroughly sour', then soaking the wollen yarn in the strained liquor, turning frequently for several days and then drying in the shade, followed by a final washing (Akeroyd & Stearn, 1985).

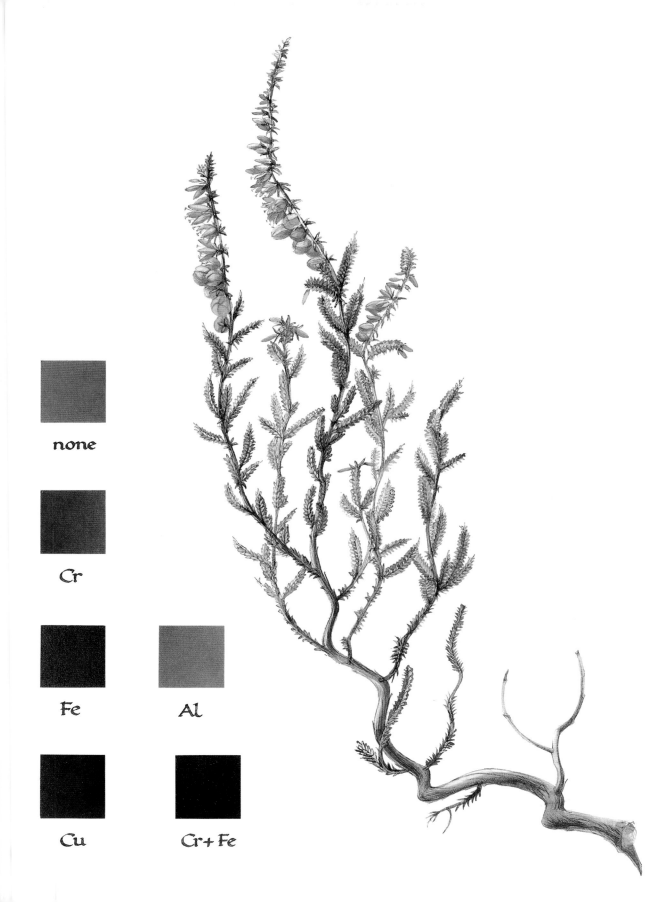

Cr

Fe

Al

Cu

Cr+ Fe

Henna

Lawsonia inermis (Loosestrife family – Lythraceae)

Henna is probably a native of the eastern part of India, but has been cultivated for many centuries throughout the Near East and North Africa. It has become naturalized in the West Indies where it is known as West Indian mignonette. Henna is a glabrous shrub or small tree of up to 5m (16ft), with a branched and twisted trunk. It has small, entire, ovate or lanceolate, privet-like leaves, in pairs or threes. The four-petalled flowers, which are very fragrant and pink, white or brownish-red, are arranged in large terminal panicles and give rise to leathery capsules about the size of a pea, which contain many angular seeds.

Henna has been used to dye the skin and hair for many centuries. Egyptian mummies have been found with their finger and toe nails, finger tips, palms and soles of the feet coloured red, orange or yellow. It was also used in ancient Egypt for colouring men's beards and moustaches, as well as for the manes and tails of horses. Around 3200 BC henna and indigo were being mixed to dye hair black. Some Jews admired this custom during the 'Captivity in Egypt' in the twelfth century BC, but its use was frowned upon by the Orthodox, who insisted that elegantly decorated female Egyptians marrying Jews must shave their heads, pare their nails and thoroughly scrub their palms and soles. Its use for fragrance is referred to in the *Song of Solomon*; its scent is said to be like that of roses, and it was still used in bouquets in Egypt early this century.

Bancroft (1813) records that the bruised leaves, moistened with water and applied to nails and finger tips, produced a dark orange-scarlet after a short time.

He says it is a substantive dye but that alum or iron would alter the colours. Henna is still used to dye hair in Europe today; a shining-red tint is given to black hair after one hour, while brown hair becomes auburn and blonde an orange-red. It was often mixed with cutch to give ginger-yellow or with indigo to give a blue-black. The dye is said to give hair a silky and lustrous sheen, and is included in shampoos for this reason. Henna is used in the Yemen, where old men can be seen with reddened beards, which indicates that they have made the pilgrimage to Mecca. Cardon (1990) records its importation as a tanning agent until about the middle of this century into Millau, a town famed for its gloves, in the Massif Central of France.

Henna is a plant of hot, dry regions and is sometimes trimmed to make a dense hedge. In Britain, it can be cultivated in a warm greenhouse and needs a compost of loam and peat. Henna is occasionally sold by nurseries specializing in what the Victorians termed 'stove plants'. Seed is also available from specialists.

The leaves are gathered and dried in summer and sold either in this form, or as a yellow-brown powder, or occasionally as a paste. The dye is substantive and the colour using unmordanted wool is light tan. Yarn mordanted with alum, chrome and copper becomes different shades of brown, while iron and tin give light brown and orange-tan. There is very little difference with alterations in the pH. The pigments present are the napthoquinones *lawsone* and the closely related *juglone*, the flavonoid *luteolin* as well as various tannins. *Lawsone* is a red, substantive dye which is very effective on the protein keratin (the main chemical component of hair).

Al   Cr   Cu

Indigo

Indigofera species (Pea family – Leguminosae)

Indigofera is a large genus of about 700 species which occur mainly in the tropics. The most widely cultivated species are: *Indigofera tinctoria* – frequently cultivated throughout the tropics, probably originating in India (a shrub up to 1m (3¼ft), with red-violet flowers crowded into spikes and cylindrical pods); *Indigofera suffruticosa* – a native of tropical America introduced into Asia and tropical West Africa (with orange to red flowers and straight pods); *Indigofera arrecta* – native to Ethiopia and is frequently cultivated in tropical Africa; introduced into Java as 'natal indigo' and from there to India where it is widely cultivated as 'java indigo'; *Indigofera argentea* – a native of Egypt and Ethiopia and cultivated throughout North Africa (is hairy, unlike the other three species, and has red-orange flowers). Several species can be grown in temperate regions and are sold by garden centres. *Indigofera heterantha* (also known as *I.gerardiana*) is the most frequently grown in Britain. It is an attractive shrub with long spikes of carmine-red flowers.

The use of indigo is mentioned in Indian manuscripts from the fourth century BC. It had been introduced by the Arabs into the Mediterranean region by the eleventh century. Marco Polo gives an interesting account of the Indian indigo industry. Indigo was imported into Britain in Elizabethan times but, as in other European countries, protests from woad growers ended in it being banned by law. It is widely used today to dye jeans, where the irregular mottling due to the rather impermanent attachment of the dye has become a cult feature.

The most important dye substance in the leaves is a colourless compound of indoxyl called *indican*, and it is extracted with hot water. Oxygen from the air converts this to *indigotin* which is blue and insoluble. Synthetic indigo is now widely used and sodium dithionite (previously named sodium hydrosulphite or 'spectralite' in the United States) is a convenient way of removing oxygen from the dye bath to make the soluble indoxyl. The chemistry is explained in greater detail under woad.

Synthetic or natural indigo should be ground with water and added to a solution of washing or caustic soda plus sodium dithionite, which is then heated to 40 to 50°C (105 to 125°F) to form a stock solution. The liquid becomes a clear yellow and small amounts are added to a pan containing enough water to cover the fibre. This dyebath is allowed to stand for thirty minutes after the addition of more sodium dithionite (to de-oxygenate the water). Wetted wool is soaked in the bath for five minutes, after which it is removed, squeezed and shaken in the air for a further five minutes. As re-oxygenation occurs, the wool turns from yellow to blue as if by magic. Successive dips and aerations are made until the right colour is obtained. Adding 1 to 3 per cent gelatine to the bath can protect the fibres from the highly alkaline solution. Fibres should be rinsed in a vinegar bath to neutralize the alkali and washed in soapy water to remove any unaltered indigo. Unused stock solution can be safely stored for many months in a deep freeze. See Grierson (1986) and Dalby & Christmas (1984) for more details. Indigo can be preserved by 'balling', a very similar process to that described for *Polygonum tinctorium* (page 70).

The indigos also contain *indirubin* in differing amounts and flavonoids derived from *kaempferol*. There is considerable variation in the amounts of pigments depending on plant age and the position of the leaves on the stems. The maximum period of production seems to be from May to August. The blue dyes are particularly light-fast when used on wool but less so on cotton. They are not attached within the fibre, but are only loosely fixed to the outside and rub off with wear.

none   none   none

Ivy

Hedera helix (Ivy family – Araliaceae)

Ivy is native to Europe and western Asia and is widely cultivated in most temperate countries of the world. It is a woody, evergreen climber which clings to walls and trees with adventitious roots. The leaves are dark green and leathery; the lower ones are palmate with triangular lobes while the upper leaves are ovate or rhomboid. The honey-scented, yellow-green flowers are borne in umbrella-like clusters on mature shoots in October to November, and the black berries mature in March to April. Flowering only occurs when the shoots have reached the tops of their supports and have ceased to climb. Ivy can tolerate very low light intensities and is often found as ground cover in quite dense woodland, as well as growing on old walls, tree trunks and in hedges.

Ivy leaves have been used to produce a black dye, but the historic importance of ivy mainly relates to a resin which was found on old ivy stems, called ivy gum. This came from southern Europe and North Africa and was the source of a red dye. The leaves were used in French country districts to help remove ink and fruit stains, although it would seem more probable that they would actually produce staining themselves.

There are a very large number of horticultural varieties, some of which can be used as indoor plants. They thrive in almost any soil and are tolerant of a wide range of light intensities, from full sun to the deepest shade. They are easy to grow from cuttings; those from flowering shoots become bushy shrubs which may grow to a metre or more and flower freely. The plant has hairs that may cause dermatitis to some sensitive skins and gloves should be worn when clearing away dead leaves; the sap can also cause irritation.

Leaves can be gathered at any time of the year and should be chopped and well-soaked in water for a day or two before use. Berries can be used either half ripe, or left on the plants till fully mature in March or April. The leaves produce yellows and greens, while the unripe berries picked in January give attractive pinkish-greys and greens. In a neutral bath, unmordanted wool becomes a pale greenish-grey, while alum, chrome and copper mordants give various greens. In an acid medium, unmordanted wool has a strong pinkish tinge, alum gives pale grey, chrome sage green and copper greyish-brown. Adding alkali gives a more yellowish tinge. The colours, however, fade considerably becoming much more yellowish and the purplish shade from unmordanted wool in an acid solution becomes a dull fawn. Skeins mordanted with iron fade much less. Mature berries picked in March give similar colours. We were, however, unable to produce the range of pinkish tones reported by some dyers.

leaves

Al

berries

none(ac)

Cr

Al

Cr (ac)

Fe

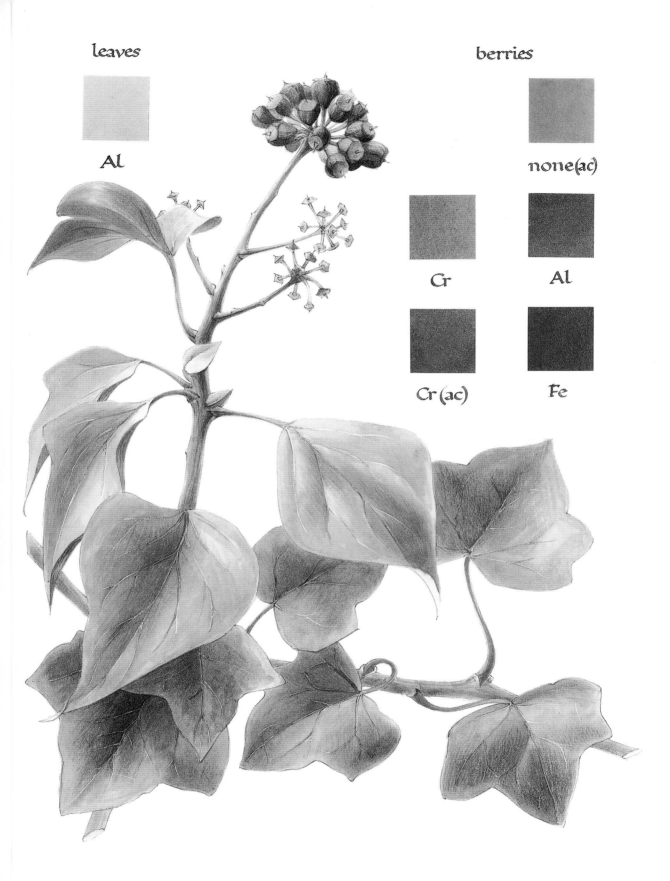

Japanese knotweed

Fallopia japonica (Dock family – Polygonaceae)

Japanese knotweed was introduced into Britain in 1825 as an ornamental and was enthusiastically planted by Victorian gardeners. It has since become extremely widespread in Europe and, to some extent, in North America. It is an aggressive weed of roadsides, railway embankments, river banks and waste places and grows up to 2m (7ft), often forming dense thickets of erect to arching stems. The coarse leaves are broadly ovate, sharply pointed and become an attractive yellow-orange in autumn. The flowers are small, dingy white, in spikes in the leaf axils. The seeds are triangular, dark and glossy.

You can buy Japanese knotweed at garden centres, but do not attempt to grow it in a small or even medium-sized garden, because it will rapidly colonize every available space. It is very difficult to eradicate; the land must be dug frequently and the roots removed and burnt, as even small pieces will grow into mature plants. Japanese knotweed should not be allowed to set fruit because it also grows readily from seed. It is classed as a noxious weed in the United States where it is banned from being planted.

A strong decoction of leaves picked in autumn gives excellent colours. The dye with alum-mordanted wool gives a strong brassy-tan, chrome an excellent tan and copper a browner tan. The colours are relatively light-fast. The pigments involved are probably anthraquinones, flavonols and tannins.

The related genus *Polygonum* has a very large number of species. *Polygonum tinctorium* is a well-known dye plant which is cultivated in Japan and southern China. It is a small biennial or annual, with thick ovate leaves and small pink flowers in branched spikes above the leaves. It was cultivated in Europe in the nineteenth century and is said to be still grown in California. Cardon (1990) notes that the use of this plant in China probably antedates the use of indigo in India. It was certainly used there at the beginning of the Middle Ages and the smell of the dye was thought to repel both serpents and mosquitoes.

Polygonum tinctorium needs well-cultivated soil and should not be sown until all risk of frost is past. It needs at least ten weeks of hot, sunny weather to produce the precursor of a blue dye. *Indirubin* is present in the plants, as well as *kaempferitrin* which is related to *indican*; the anthraquinone *polygonin* and several flavonols are also present. The plants are also rich in flavonoids, which give the blue dyes a slightly green tint.

Dyeing methods are similar to those used for indigo. To preserve the leaves of indigo-bearing plants the 'balling' process is used. This starts with crushing the leaves, which are heaped into piles and the sap allowed to drain. The pulp is kneaded by hand into balls of between two and six inches, and dried on trays. Once dry, balls can be used immediately or they can be 'couched'. This involves grinding the balls to powder, which is wetted and fermented for up to nine weeks, turning the smelly mass and sprinkling it with water several times. Eventually it forms a clay-like paste which can be dried for later use.

Other polygonums which contain a blue dye include the common knotgrass (*P.aviculare*) which is cultivated in Japan for the dye. It is commonly found along paths and roadsides in Europe and has been introduced into the United States and many other temperate parts of the world. It is a small, more or less prostrate annual with leaves of two sizes and the very small pink or white flowers are found at the bases of the leaves. It may be worth dyers experimenting with this plant after a long, hot summer. Fermented stems of buckwheat (*Fagopyrum esculentum*) are said also to give a blue dye.

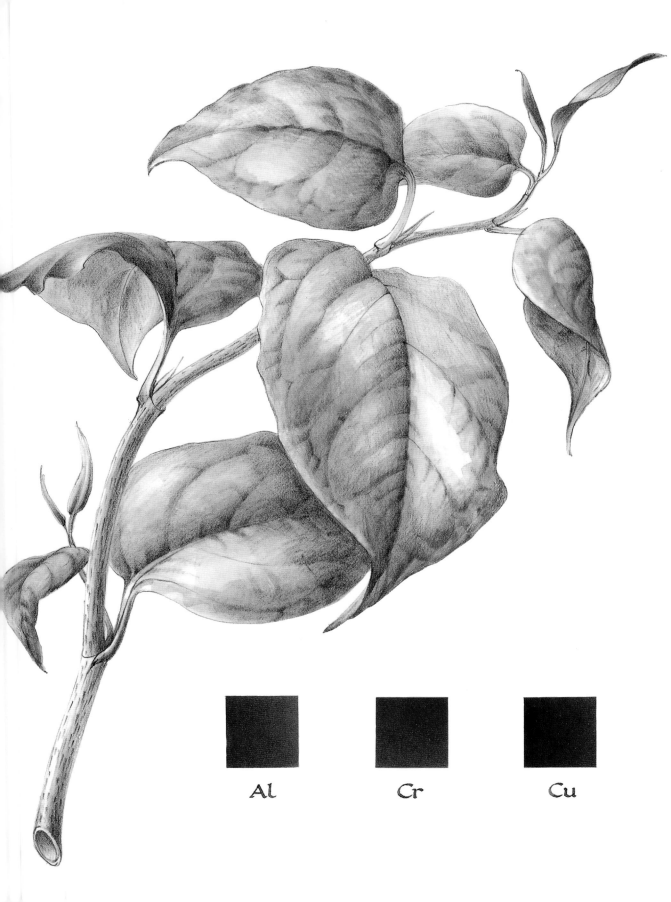

Al   Cr   Cu

Lady's bedstraw

Galium verum (Madder family – Rubiaceae)

A native of Europe and western Asia, Lady's bedstraw is a low-growing perennial and has weak, slightly four-angled stems, blackening when dried. The yellow flowers are in leafy panicles, the fruit is black and two lobed, and the plant has a strong scent of new-mown hay. It is found on all but the most acid soils, often on sandy banks, pastures and on fixed sand dunes, and is widespread in Britain.

Roots of Lady's bedstraw were used in Scotland in the production of tartans. The plants were easily gathered from sandy pastures and dunes, but collecting caused so much erosion, particularly on the rich machair soils, that an act of Parliament in 1695 prohibited gathering roots from these places. However, the dye was so highly prized that raids took place at night as late as 1915. Lady's bedstraw was used to colour cheese in several parts of the British Isles and dairy maids in the Midlands are said to have used it to colour their hair. Culpeper in his *British Herbal* of 1652 says, 'the decoction of the herb and flower is good to bathe the feet of travellers and lacquies, whose long running causeth weariness and stiffness in their sinews and joints'.

Yellow dyes are present in the flowering tops, which can be collected sparingly from the wild. A red dye is present in the roots but, in line with the requirements of the law, these should not be collected without the permission of the landowner. The plant can be grown easily from seed, which must be gathered when ripe in August or September, and then sown at once in seed compost. Seed can be bought and a few nurseries stock the plants. Lady's bedstraw is best planted in very deep, loose sandy soil with the addition of a little lime, but should not be planted out near a rock garden as the roots are very invasive. Plants grown for the red dye should be left for three years before harvesting. The best dye-containing roots grow near the surface and can become 1cm (⅓in.) thick. Thinner roots also contain pigments but they produce yellower shades.

Flowering tops should be gathered in July or August when the plants are in full flower, while roots should be dug in the autumn and are best used after slow and thorough drying. The tops are chopped and simmered in the usual way and produce a range of yellows and greens which are moderately resistant to fading. Unmordanted and alum-mordanted wool give similar shades of yellow, while chrome and copper give orange and yellow-green respectively. Copper-mordanted wool fades very slightly. Adding iron to the dye bath produces khaki, brownish-khaki and olive-green with alum, chrome and copper respectively.

The roots contain various anthraquinone pigments, including *alizarin, purpuroxanthin, rubiadin, purpurin* and *lucidin*. The colours are coral-reds and salmon-pinks and are resistant to light and washing. Several additions to the dye bath to enhance the colours have been suggested, such as malt, yoghurt and various enzymes, but the shades do not alter much with changes in pH. The roots are treated in the same way as madder and if the dye bath exceeds 60°C (140°F), the clear reds become much browner. Dried roots are also very susceptible to deterioration by the absorption of water. Grierson (in Buchanan, 1990) describes a process which gives a good red colour on wool, during which the roots are soaked, simmered and then re-simmered at a temperature higher than is often recommended for madder and related plants.

Other species of *Galium* contain numerous similar anthraquinones in their roots. They include heath bedstraw (*G.saxatile*), a slender white-flowered plant of acid soils; hedge bedstraw (*G.mollugo*), a much larger, white-flowered plant of hedges and similar habitats on calcareous and other nutrient-rich soils; sweet woodruff (*G.odoratum*), a white-flowered plant of damp and calcareous woodlands, which smells of new-mown hay and was used by the Elizabethans to scent their linen; and goosegrass (*G.aparine*) which is a troublesome garden weed.

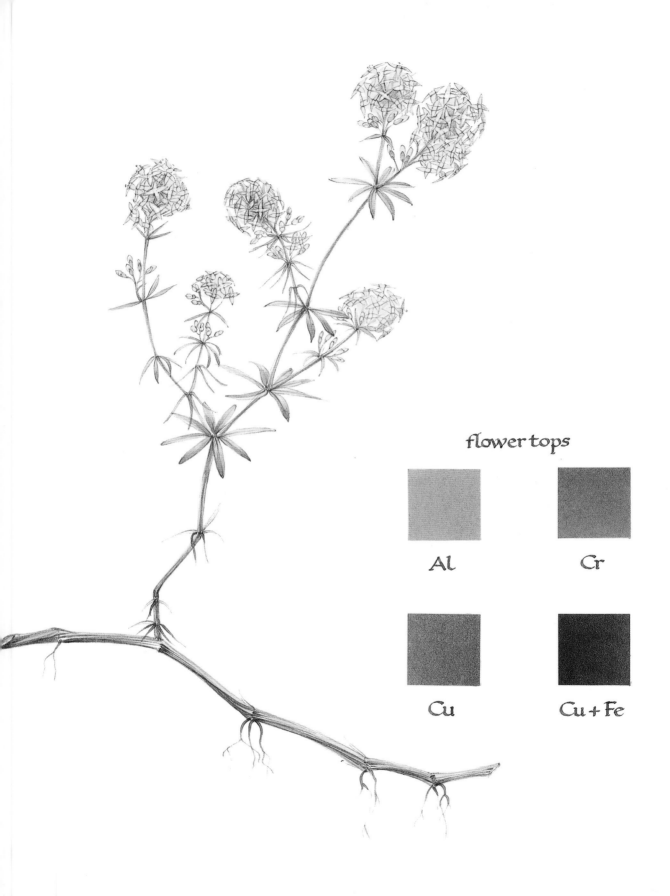

flower tops

Al

Cr

Cu

Cu + Fe

Logwood

Haematoxylon campechianum (Pea family – Leguminosae)

Logwood (Campeachy wood) is a native of tropical Central America and northern South America, and is cultivated in these regions as well as the West Indies, Brazil, India, Ghana and Madagascar. It is a tree of up to 15m (50ft), with a reddish trunk that becomes twisted and gnarled with age, and is covered with thorns. The leaves are pinnate, sometimes bipinnate at the base, with two to four pairs of obcordate leaflets. The numerous sweet-scented, yellow flowers are borne in racemes about as long as the leaves. The pod is flat and wing-like, 35 to 50cm (14–20ins) long with one to four oblong seeds. Although it is easily grown from seed, it needs a tropical climate to flourish.

Logwood was discovered in the Mexican state of Campeche by Spanish conquistadores in the sixteenth century and was probably used by the Aztecs before this. Its export to Spain was quickly followed by importation into many European centres of trade; it was imported into England by the middle of the sixteenth century, and in 1581 a law prohibiting its use was passed to protect the woad growers, but it was still used illicitly. Methods of mordanting were primitive and obviously not very effective, and this probably influenced this blue dye's reputation for poor light-fastness. However, it was used more as a black dye and was imported in increasing quantities after the repeal of the law in 1661. As it increased in importance it was cultivated in Jamaica and Belize (formerly British Honduras), which is said to have owed its origin as a colony and subsequent prosperity to logwood. Squabbles between the Spanish and the British over the exploitation of the wood became increasingly bitter, involving sea battles, pirates and buccaneers. Ponting (1980) gives an account of this history. The trees flourished in the West Indies, and Jamaica exported large quantities of the logs to Europe and North America. Logwood still remains an important dye substance, and Cardon (1990) states that as late as 1943, more than 70,000 tonnes were being used annually worldwide. It was used as a black dye on silk until very recently and is still used to dye nylon and as a very important stain for biological work.

The logwood dye is concentrated in the heartwood. It is available from dye suppliers as dark-coloured chips. These are usually soaked for at least twenty-four hours before simmering for half an hour or more. Thirty grams (1oz) of logwood chips are sufficient to dye 100g (3½oz) of wool to a deep shade. The chips can be reboiled to give progressively lighter shades. Logwood can be used to give several different colours, including blue, violet-purple and black. Our own experiments produced an interesting range of colours from powdered, unsoaked chippings, over which boiling water was poured. Quick boiling gave pale blue on unmordanted wool, pinkish-blue with chrome and Prussian blue with copper. Adding washing soda to the dye bath gave much darker, duller colours, while adding acid gave different results. Unmordanted wool became yellowish-buff coloured while alum, chrome and copper gave tan, purplish-brown and black respectively. Iron gave grey with unmordanted wool and blacks with all three mordanted skeins. Adding a small amount of methylated spirits to the wood chips, then steeping skeins in the extract, produced strong colours without heating; a good violet-blue was produced on the alum-mordanted wool, violet with chrome, blue-black with copper and a soft mauve with tin. Cardon (1990) recommends logwood at 30 per cent of the weight of the wool, with the same proportion of calcium acetate, to give an indigo blue; the shade can be altered with an after-bath of tin and changes in pH. Black is usually obtained by a similar method with wool premordanted with chrome or copper. The dyes present include tannins, which are important in the production of black. The blue in the heartwood is *haematoxylin*, which is closely related to the flavonoids. When the tree is cut, the yellow wood turns black as the *haematoxylin* is oxidized into *haematin*.

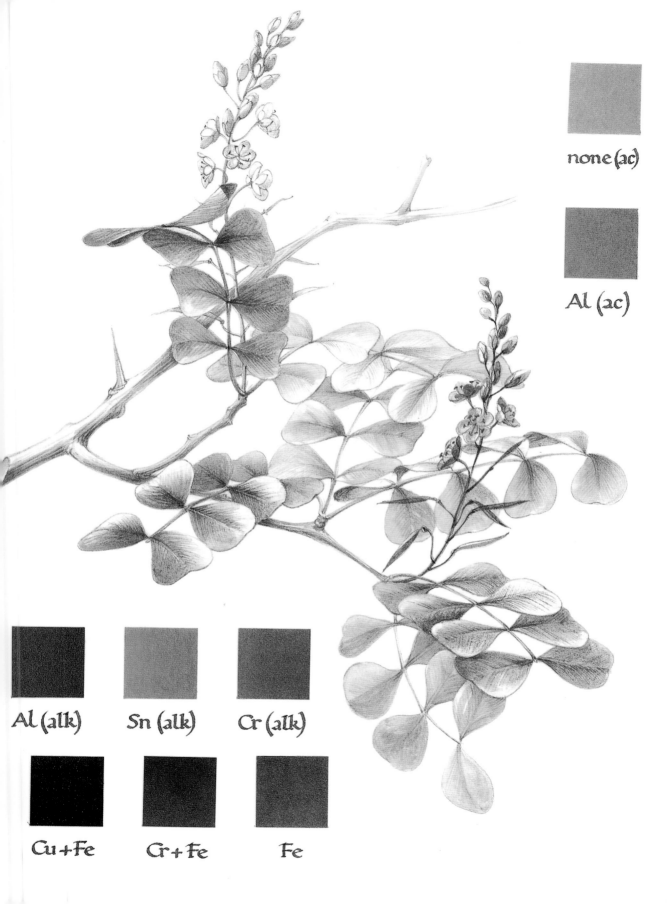

none (ac)

Al (ac)

Al (alk)   Sn (alk)   Cr (alk)

Cu + Fe   Cr + Fe   Fe

Madder

Rubia tinctoria (Madder family – Rubiaceae)

Madder is native to western and central Asia and doubtfully so in the eastern Mediterranean. It has become naturalized in many areas of central and southern Europe and occurs occasionally in the British Isles. The stems are up to 1m (3¼ft) long, bear downward-pointing prickles, and straggle along the ground or scramble through other vegetation. The lanceolate or elliptical leaves are borne in whorls of 4 to 6, while the terminal panicles are composed of numerous, small, pale yellowish flowers with four petals. The ripe berries are black. When naturalized it is often found in hedges, thickets or waste places.

Madder has been used for many centuries and cotton textiles dyed with it, dated around 3000 BC, are known from the Indus civilization. It has been found on linen from the Nile Valley tombs and is mentioned in the Bible. It was prized in ancient Greece and Rome and cultivated in Italy and the Near East. In the Middle Ages, madder became of great importance in Europe and was grown extensively in France and Germany. Between the sixteenth and eighteenth centuries, production had become almost a Dutch monopoly and, from there, it was even exported to India for the expanding cotton industry.

The strong, almost fadeless cotton dye known as turkey red was developed in India and spread from there to Turkey. It involved about twenty separate processes using blood, oil and rancid fat, as well as charcoal, cow, sheep and dog dung and the liquid contents of animals' stomachs. Villages where the process was carried out were, not surprisingly, occupied only by the dyers and their families. Greek workers skilled in the techniques were brought to France in 1747. Industrial espionage resulted in the secrets of the process being stolen by spies sent from Holland and England and a sanitized version of the process was adopted in Manchester by 1784. Madder was used for soldiers' uniforms and for hunting 'pink' coats. The use of madder declined abruptly following the development of synthetic *alizarin* in 1826.

Madder can be bought from some garden centres or grown from seed. It prefers well-drained, well-cultivated soil with a little lime and can even thrive on almost pure chalk. The stems are brittle and need supports to allow the plants to flourish. Madder should be grown for at least three years before harvesting. The roots are lifted in the autumn and portions retained for replanting. As might be expected, the largest roots contain the most pigment; smaller ones are usable but can give rather different colours.

After lifting, remove the remaining soil and spread the roots out to dry in a very slow oven. Do not overheat them otherwise the colours will be less intense. Care must also be taken to keep them very dry because they easily absorb moisture, with subsequent deterioration in quality. The outer parts of the roots and old stems are pounded to produce an inferior dye called mull, once used to produce browns. The best-quality madder is called crap (krappe or crop); it comes from the central part of the roots and is ground into a powder before use. Home-grown madder roots can be used whole or pounded, if separation into the different portions is impracticable.

Dyeing with madder must take place slowly. Premordanted wool is placed in the dye bath with 30 to 40 per cent of its weight of powdered madder. Chalk or slaked lime is added, particularly in areas with soft water. The temperature is then slowly raised to 85°C (185°F), which must not be exceeded nor continued for more than two hours. If the temperature falls, part of the dye may be precipitated and lost; if it is exceeded the colours become dull and brown. Washing in hot, soapy water helps to eliminate any yellow tints present. Scarlet is produced by adding tin to the dye bath for the last fifteen minutes for wool premordanted with alum, or by premordanting with alum and tin. The colours, which vary with the quality of the madder and the mordants used, range from pale apricot, to pink, red, orange and purple. The most important pigments in madder roots are the anthraquinones *alizarin*, *rubiadin*, *purpurin*, *pseudopurpurin*, and *mungistin*.

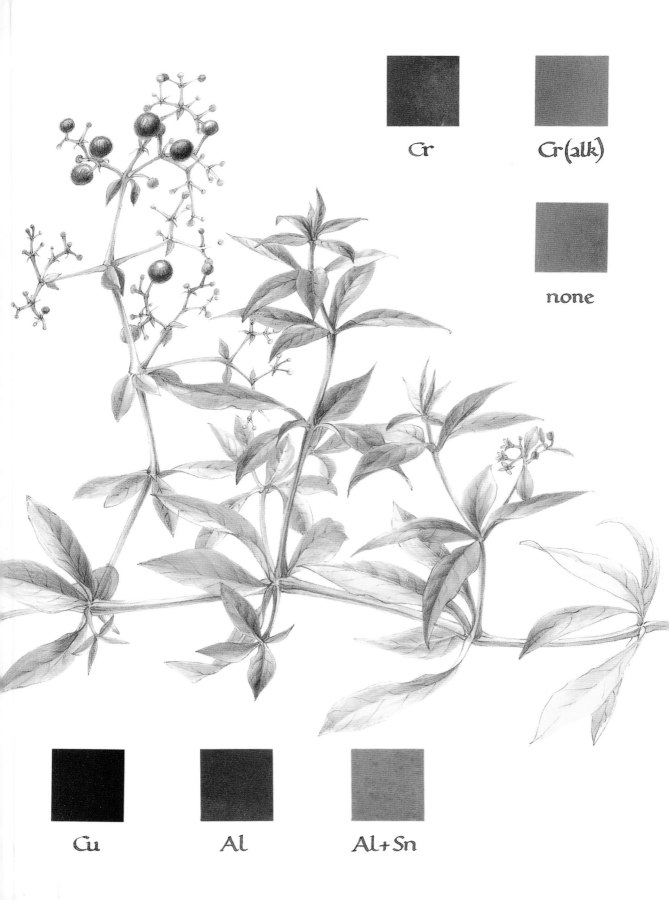

Cr

Cr(alk)

Cu

Al

Al+Sn

Mahonia

Mahonia japonica (Barberry family – Berberidaceae)

Despite its botanical name, this species is a native of China, though long cultivated in Japan. It is an erect, branched, evergreen shrub of up to 3m (10ft) with leathery leaves up to 40cm (16in.) long and with 13 to 19 spiny leaflets. Fragrant yellow flowers are borne in drooping clusters and are produced from December to March, followed by bluish-purple berries.

This attractive shrub is available from garden centres and is sometimes sold under the name of *Mahonia bealii*. A lower-growing variety with broader leaflets and erect flowers called 'Hiemalis' is also sold, together with various hybrids with *M.lomariifolia*, such as 'Charity', 'Winter Sun' and 'Lionel Fortescue'. Mahonias will flourish in most well-cultivated soils and can be propagated from seed sown soon after ripening, or from cuttings taken just below a leaf bud in autumn or winter.

The wood of *M.japonica* is yellowish and contains some pigments, but the bark gives better results. This should be chopped and pre-soaked for twenty-four hours before use. In a neutral bath, unmordanted and alum-mordanted wool become greenish-yellow; chrome gives a more greenish colour and copper produces olive-green. Adding iron to the dye bath gives deep sage and olive-greens. The leaves, chopped and left in water to ferment for seven to ten days, give colours that are much more yellow than those from the bark. The best results come from wool that has been warmed to about 50°C (122°F) with the fermenting leaves, then cooled and allowed to stand for about forty-eight hours. Unmordanted wool takes up very little colour, but alum-mordanted wool becomes a pleasing lemon-yellow, while chrome and copper give greyish-yellows. Boiling the fermented leaf liquor for thirty minutes produces a pleasant brassy colour with chrome. The pigments present are probably similar to those of other species of *Mahonia*, of which the isoquinoline alkaloid *berberine* is the most important. The berries contain anthocyanins.

Mahonia aquifolium, the Oregon grape, is a native of western North America. It is naturalized in some places in Europe and is widely cultivated in parks and gardens. These shrubs have leaves with 5 to 9 leaflets which resemble holly, small yellow flowers and blue-black berries. The bright orange roots and the stems produce fairly permanent golden-yellow and brownish-reds. The dyes present are the alkaloids *berberine*, *berbamine* and *oxyacanthine*. The berries are best in an acid solution, when they produce violet-blues. The pigments present are anthocyanins. The Navaho Indians of the southwest United States use Oregon grape for dyeing, but their use of vegetable dyes, with the possible exception of indigo and goldenrod, was introduced as late as the 1920s and is not of historical importance.

Several species of *Berberis*, another genus of the barberry family, have been widely used. *Berberis vulgaris*, from Europe and western Asia, and widely naturalized in England, is a spiny shrub of up to 2.5m (8ft) with yellow flowers and red berries and usually occurs on calcareous soils. *Berberis laurina* produces a yellow aqueous extract used by herdsmen in the state of Rio Grande, Brazil, to dye their saddle blankets a golden tint. Species of *Berberis* can be bought from garden centres and thrive in well-cultivated soils. They are not available in many parts of the United States, because they act as intermediate host plants to fungi that attack cereal crops. The pigments present in the roots, stems and leaves are alkaloids, the most important being *berberine*. *Berberis vulgaris* was mentioned as a dye plant in a fourteenth-century Austrian manuscript. Linnaeus records the use of *B.dumetorum* as a yellow dye while Bancroft (1813) notes *Berberis* as the source of a substantive dye on wool, but was not impressed by its permanence either to soap or exposure to sunlight. Mairet (1916) states its use was principally for silk and that it was used by nomadic Anatolian tribes for yellow-brown dyes.

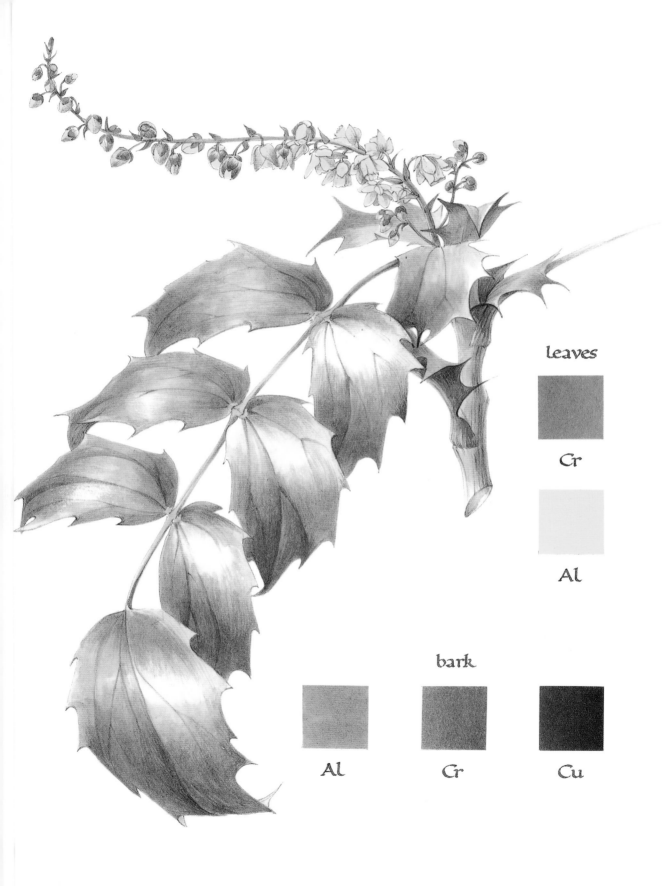

leaves

Cr

Al

bark

Al

Cr

Cu

Munjeet

Rubia cordifolia (Madder family – Rubiaceae)

Munjeet comes from the mountainous regions of Asia, from the Himalayas to Japan and also occurs in Sri Lanka, Malaysia and tropical Africa. It is widely grown throughout tropical Asia for its excellent red dye.

Munjeet is a trailing or scrambling plant with whorls of stalked, heart-shaped leaves. The very small, whitish flowers are massed in panicles above the leaves and the berries are black. Subtropical conditions are needed to grow the plant. Munjeet was an important dye for the Asian cotton industry as the use of a native plant was obviously much more economical than imported products. It is still used in Nepal today in areas where interest in the use of native dyes is being fostered. The whole plant, not just the roots, is dried and used to dye various fibres reds and pinks (Hecht, 1989).

Colours produced on wool in a dye bath held at a temperature of 85°C (185°F) are remarkably light-fast, although unmordanted wool is less permanently dyed. Without a mordant, strong orange results; alum gives an excellent red; chrome and copper give reds with slightly brownish tones. With the addition of tin, the alum-mordanted wool becomes a bright orange-red. A bath containing a small amount of roots or an 'exhaust bath', i.e. one that has already been used for dyeing but still has some pigment left, can give pretty shades of apricot and pale brownish-orange.

Munjistin, an anthraquinone, is found in the roots of munjeet, together with the pigments *purpurin*, *pseudopurpurin* and *purpuroxanthin*. These are also found in madder.

Wild madder (*Rubia peregrina*), a native of Europe, Turkey and North Africa, occurs in the coastal counties of southern England and to a lesser extent in the coastal areas of Wales and Ireland. It is a trailing or scrambling evergreen plant and the lower part of the stems are persistent and woody. The downward-pointing prickles of the stems assist the plant to ascend through other vegetation. There are 4 to 6 leaves in each whorl which also have prickles on the underside of the midribs. The flowers are small, yellowish-green and in leafy panicles, while the ripe berries are black.

The roots of this species are smaller than those of madder (*R.tinctoria*) (page 77), but are used in the same way. It does not seem to be available commercially as plants or seed, but it can be cultivated from seed collected sparingly in the wild. Wild madder should be treated in the same way as madder, although you need more roots for a good colour. The pigments present include *purpurin*, but not *alizarin*. This plant was used for Coptic textiles of the sixth century AD, together with an indigo blue to produce a violet colour known as Egyptian purple (Cardon, 1990).

Cochineal is another dye which gives strong and very permanent reds and was particularly prized for its ability to produce a vivid scarlet. It is formed by females of *Dactylopius coccus*, a plant bug which feeds almost exclusively on cacti of the genus *Opuntia*. It was bred and used in Mexico and brought back from there after the Spanish conquest. It rapidly supplanted kermes, another insect dye, which had been used from very early times for scarlet. It has been identified as a dye on fabric dated as early as the first century AD in Peru. The pigment present is *carminic acid*, which, like other anthraquinones, turns from red to blue with changes in pH. Its use for scarlet involves mordanting with tin in the presence of oxalic acid and sometimes turmeric is used with it to enhance the orange tones of the scarlet and to counteract the tendency to turn bluish in soapy (alkaline) water.

Cr

Cu

Al+Sn

Al

Nettle

Urtica dioica (Nettle family – Urticaceae)

Stinging nettles are natives of Europe and temperate parts of Asia, but they are widely introduced elsewhere. They are found on cultivated ground, waste places, hedgerows and particularly near buildings. Nettles often indicate the presence of an old building which may no longer be apparent. Occasionally, nettles are found without stinging hairs, which would be ideal for dyers wanting to grow the plants. Nettles are also grown by those with their own garden nature reserves; they are food plants of the caterpillars of peacock and small tortoiseshell butterflies. The plant can be eaten as spinach or used as an ingredient for an excellent soup. Nettles were sometimes used in Scotland to produce yellows and dark greens for Harris tweeds. Grierson (1986) gives a reference to its use in Ireland.

The ubiquitous stinging nettle is a perennial herb of up to 1.5m (4½ft), rarely 2.5m (8ft). The erect shoots arise in spring and summer from creeping stems which root at the nodes, forming tough, spreading yellow roots. The leaves are 4 to 8cm (1½ to 3¼ins), in opposite pairs, with deeply toothed margins and are heart-shaped at the base. The small, greenish flowers are unisexual and arranged in catkins, usually on different plants. The male catkins are spreading, while the females are pendulous.

The yellow roots, though small in diameter, can easily be pulled out of loosened soil. They are said to produce a good yellow dye with alum. It is, however, more usual to dye with the young leafy tops, which can be chopped and boiled in the usual way.

The colours produced vary greatly with the season and probably with the habitat. Yellows, yellow-greens, grey-greens and golds have been reported by various authors. Our own experiments have been made with tops picked in July from plants growing in partial shade. Unmordanted wool gave a very pale cream colour, alum a pale fawn, chrome a tan and copper a greenish-brown. These colours faded slightly after exposure to partial sunshine for three weeks. Wools that were first dyed with nettles using alum, chrome or copper, and subsequently 'saddened' with iron, gave strong dark browns, or a grey-brown colour with unmordanted wool. These colours were very light-fast when tested in the same way. The pigments present include *lutein* and other carotenoids including *cucurbitin*.

The stings of nettles have pointed tips which penetrate the skin and break off to allow the irritant fluid in the hair to enter the skin. Although the stings of the common European nettle are unpleasant, their effects are not long-lasting and many people think that the smaller and less familiar annual nettle (*U.urens*), a weed of arable fields and vegetable gardens, is worse. A Himalayan species, (*U.crenulata*) produces violent inflammation, the pain of which sometimes spreads throughout the body, lasts for many hours and only disappears after nine or ten days. In Europe, since the Bronze Age, the fibres in nettle stems have been used for spinning into a linen-like thread. In Germany, cloth woven from these threads, known as *nessel tuch*, could be bleached as white as linen, and even as late as the First World War nettle fibre was used to supplement cotton supplies. Ramie (*Boehmeria nivea*) is another member of the nettle family which is cultivated mainly in China and Formosa to produce a white, lustrous, non-elastic yarn used mainly for heavy industrial fabrics (Baines, 1989), although it is possible to spin a yarn from it nearly as fine as silk, which in China is sometimes used for lace-making. A Himalayan nettle, allo, (*Girardinia diversifolia*) has been used traditionally in Nepal and its fibres are currently being developed for wider use (Dunsmore 1988a & b).

Al

Fe

Cu

Cr

Onion

Allium cepa (Lily family – Liliaceae)

The onion used in kitchens throughout the world is unknown in the wild. It has been cultivated for at least 3000 years and is probably derived from a central Asian species *Allium oschaninii*. Onions, leeks and garlic are referred to in the Bible and the Israelites are described as yearning for them after the flight from Egypt. There is considerable variation between the numerous cultivated varieties, but the bulbs are often flattened-globose and covered with several layers of papery skins, known as the tunic. They are most familiar in Europe as the 'Egyptian' onion which has a yellow-brown skin. The leaves are almost tubular and up to 40cm (16ins) long. They are basal in the first year but, in the second year, sheath the lower part of the flowering stem. The flowers are star shaped and are borne in an umbel. The petals are white with a green stripe.

Detailed instructions for the cultivation of onions can be found in any good gardening manual. Onions are extremely sensitive to day length and temperature. Roots and leaves are formed when daylight does not exceed 15 to 16 hours, while the fleshy bulb scales are produced only during the longer summer days. Onions prefer good medium-to-light soils and do not produce heavy crops on dry sands, thin chalk soils or sticky clays.

The production of onions with the brown-coloured dry tunic scales is best left to the professionals. Dry scales can readily be obtained from greengrocers, as there are always plenty of skins left after a batch of onions has been sold. Although the skins contain quite a lot of colour, they weigh very little and a lot are needed for dyeing. Thoroughly dry skins can be stored in paper bags for several years.

Quite strong colours can be obtained using unmordanted wool, although they may not be very light-fast. The dye is said to be substantive but if mordants are used different colours can be produced. A strong bath of approximately twice the weight of skins to wool gives a pinkish-brown with no mordant, bright orange with alum and tin, orange-tan with chrome and darker brown with copper. A bath of approximately one fifth the strength of the former gives pale yellow with no mordant, bright yellow with alum, yellow-orange with tin, pale khaki with chrome and brown with copper. The green parts of onion plants can be used, but give paler shades than those from dried skins; the wool becomes very smelly and it requires several washings and airings to remove the odour. The most important dye substances in onions are flavonoids, the flavonol *quercetin* is present in considerable quantity, as are several related compounds that have not yet been identified with certainty, as well as the anthocyanin *paeonidin* (known to the European food industry as E.163(e)); *protocatechic acid* and some tannins are also present. Red or purple onion skins (which may give slightly different colours in dyeing when compared to the more usual yellow-brown onions) contain anthocyanins.

The leek (*A.schoenoprasum*) is also a source of dyes similar to those of the green parts of onions. The green leaves should be chopped and used in the same way as onion leaves. The colours produced are rather dingy yellows and dull browns. A wild onion, ramsons (*A.ursinum*), is often abundant in damp woodland. It has a small bulb and a pair of large, ovate green leaves up to 20cm (8ins) long by 8cm (3¼ins) wide. The inflorescence stems are triangular and bear an umbel of about twenty white, star-shaped flowers. It is a native of much of Europe, the Caucasus and Asia Minor. It is only suitable for cultivation in a large woodland garden. However, it is often so abundant that it could be safely gathered, in moderation, from the wild. The colours produced are dull yellows and browns.

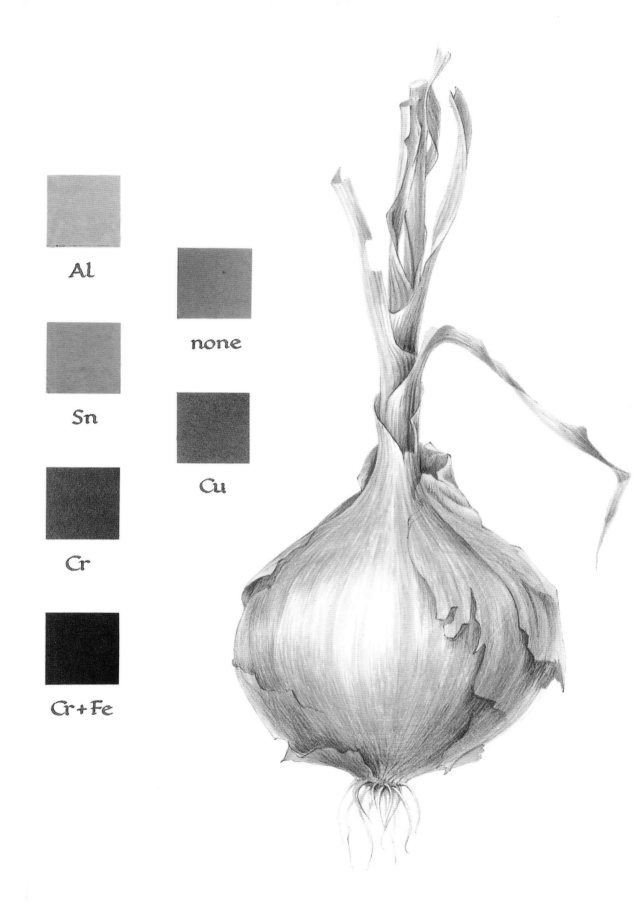

Al

Sn

Cu

Cr

Cr+Fe

Osage-orange

Maclura pomifera (Mulberry family – Moraceae)

The Osage-orange is native to the south-central United States and is widely planted in the eastern and occasionally in the northeastern states. It is a lowland tree growing best in the deep, rich soils of river valleys, but also tolerates a wide range of soils. It is often planted as a hedge and pruned into a dense, thorny barrier. The name Osage derives from the Osage Indians, who were said to use the trees for canoes and bows and occasionally in basketry.

Osage-orange is a medium-sized tree of up to 20m (65ft) and the trunk may be up to a metre (3¼ft) in diameter, with thick orange-brown bark. The branchlets bear stout green thorns that become light orange-brown at maturity; the leaves, on slender stalks, are up to 12cm by 7cm (4¾ by 2¾ins), broadest below the middle and tapering to a heart-shaped base. The flowers, borne in May or June, are either male or female, and these are usually found on separate trees. The male flowers are grouped in dense, stalked, hanging clusters up to nearly 4cm (1½ins) long, while the female flowers are arranged in dense, shortly-stalked globular heads up to 2.5cm (1in.) in diameter. The fruit are large and fleshy, up to 14cm (5½ins) in diameter and covered with small greenish-yellow projections on the surface, the whole resembling a green, warty orange. Osage-orange is available from specialist nurseries in Europe; it is hardy and is particularly successful on thin chalky soils.

The bark and the yellow-coloured wood are the parts of the plant most frequently used for dyeing. The roots, which are bright orange, are also said to contain a dye substance, as do the fruit. The bark and wood can be used fresh, but the dried, chipped wood is readily available in Europe from dye suppliers and an extract is also available in the United States. The colours produced from the wood of quite small branches are very similar to those from commercially prepared wood chips, but unmordanted wool is more yellow and less orange in tone while copper-mordanted wool is a yellowish rather than a brownish colour. Alum-mordanted wool becomes a yellow-orange and chromed wool a brownish-orange with both fresh and commercially prepared wood. The dye present in the wood is *morin*, a flavonol, while the fruit contains two isoflavones called *osajin* and *pomiferin* which are confined to this species.

Another member of the Mulberry family that is of interest is the paper mulberry tree (*Broussonetia papyrifera*). This has been traditionally used in Polynesian communities for the manufacture of tapa- or bark-cloth for garments and blankets etc. It is prepared from sheets of the inner bark of the tree, which are soaked for several days in shallow containers of water. The wet bark is then beaten on smooth stones with a tapa beater (a special form of grooved wooden mallet), which is of great mystical significance to Polynesian women. During the beating process, the bark is gradually turned into a stiff paper-like sheet, which can be thickened by beating several sheets together to form a composite fabric. Bark cloth was of great importance throughout the Pacific, as it was a quicker method for cloth production than finger weaving, looms being unknown before the arrival of the early explorers. Although, of course, western textiles are now readily available throughout Polynesia, tapa cloth is still made for ceremonial purposes and for sale, painted with traditional designs, to tourists.

Al   Cr   Cu

Persian berries

Rhamnus species (Buckthorn family – Rhamnaceae)

Several species of *Rhamnus* have been used over the centuries, and the names for the dyes came mainly from the regions where the plants were collected. French berries (graines d'Avignon), Spanish, Italian, Levant and Hungarian berries were collected from several species. They were mixtures of *R.saxatilis* (subspecies *saxatilis* and *tinctorius*). (*Rhamnus infectorius*, which is often mentioned in the literature, is now regarded as a synonym of subspecies *saxatilis*.) *Rhamnus catharticus* and *R.alaternus* were also gathered. Persian berries were mixtures of *R.saxatilis* and other species. *Rhamnus alaternus* is a small, spineless evergreen tree with oval or lance-shaped leaves up to 6cm (2½ins) and the small, yellowish flowers give rise to reddish berries which turn black when ripe. *Rhamnus catharticus* is a small, rather thorny tree, often found in hedgerows, scrub and woodlands on chalky soil in the British Isles and many continental European countries. *Rhamnus saxatilis* is a procumbent to erect shrub with larger black berries.

In the Middle Ages, the clothing, and especially the hats, of Jews was dyed yellow to distinguish them from Christians, a reflection of the yellow turbans or caps which the Persians made Christians and Jews wear in the ninth century and later. These hats were often dyed with Persian berries. The berries were also used in calico printing, often as the yellow component of the greens. In China in the nineteenth century, a beautiful green dye was produced, known as Vert-de-Chine. It was made without the use of indigo from the bark of *R.utilis*. It was very expensive and used only on the finest of fabrics. Later, French chemists produced a similar colour using *R.catharticus* bark by boiling it with a large amount of alkali. The pigment is thought to have been *rhamnicoside*. An extremely costly blue dye known as lokau was made in China by a very laborious process from the bark of *R.dahurica* or *R.saxatilis*, used solely for export, particularly for the French silk industry (Grierson, 1989).

You can buy plants of *R.alaternus* and *R.catharticus* from specialist nurseries and they will grow in many types of soil. *Rhamnus alaternus*, however, is not recommended for cold inland areas. It is well suited to maritime and urban areas. The unripe berries should be picked when fully grown, but before they go black, in July or August.

The colours from imported Persian berries are pale yellow with unmordanted wool, yellow-orange with alum, brassy-orange with chrome and brownish with copper. However, colours vary when different methods of soaking the berries are used. For example, we have used Cardon's (1990) method heating the berries slowly up to 37°C (98°F) (and holding them at that temperature for an hour or more), then allowing them to come to the boil slowly. Our trial gave a brilliant orange-yellow with tin. If, however, the berries are plunged into boiling water and then simmered about an hour with the wool skeins, the colour with tin is a strong lemon yellow. The different results are caused by enzymes which are destroyed by boiling water, while 37°C (98°F) is the optimum temperature for their reactions.

The colours from *Rhamnus* berries, particularly with alum, have a tendency to become redder with age and are very light-sensitive. Methods to prevent this reddening are given by Cardon (1990), using very slow simmering with wool premordanted with tin and using copper as an after-mordant. *Rhamnus catharticus* berries together with leafy twigs give pleasing shades of brown. Unmordanted wool gives a pale fawn which fades very little, alum-mordanted wool loses much of its colour in sunlight, while chrome- and copper-mordanted wools fade very slightly. Adding iron gives a deep khaki colour on alum-mordanted wool which appears to fade very little.

The pigments present in the berries are flavonols, in particular *quercetin*, *kaempferol*, *rhamnetin* and *rhamnocetrin*, as well as anthraquinones, the most important of which is *emodin*. Some of the enzymes present, especially *rhamnase*, modify some of the pigments. The bark of *R.catharticus* contains the anthraquinones *emodin*, *frangularoside* and *rhamnicoside*; that of *R.alaternus* contains *emodin* and *alaternine*.

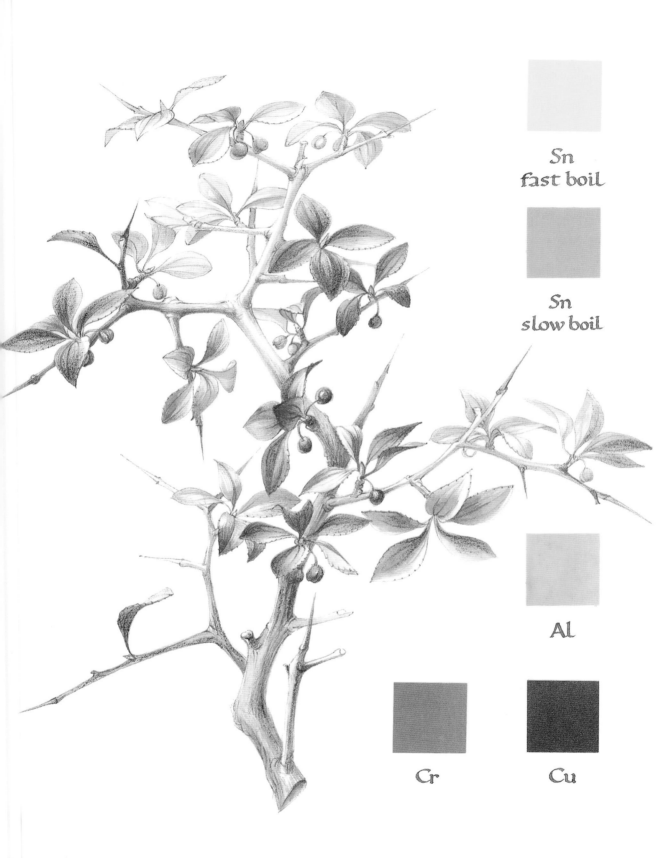

Sn
fast boil

Sn
slow boil

Al

Cr

Cu

Pokeweed

Phytolacca americana (Pokeweed family – Phytolaccaceae)

Pokeweed is a native of North America, growing wild from Canada to Texas and Mexico. It has been cultivated in Europe as an ornamental and for the red dye found in the berries. In its native range, it is found on rich soils in thickets and the borders of woods and on waste ground, roadsides and railway tracks. In Europe, it occurs in similar places, being widely naturalized in the south, and occurs occasionally in the British Isles. It is a large perennial herb, somewhat woody at the base, with red, branched stems 1 to 3m (3 to 10ft) tall bearing stalked, broadly-ovate leaves with violet veins, up to 30cm (12ins) long. The small, rather star-shaped flowers are greenish or pinkish and are borne in erect and spreading spikes, giving rise to purple-black, ten-lobed berries about 10mm ($\frac{3}{8}$in.) in diameter. Pokeweed is sometimes known as the 'red-ink plant' because the colour from the berries is sometimes used as an ink or paint. It is used to colour wine, pork products, confectionery and paper. As it is a plant product, food manufacturers can legitimately claim that they use only 'natural colouring'.

Pokeweed can be easily grown in most soils. It is suitable for a medium-to-large garden and tends to spread by seeds sown by birds. The roots, purple parts of the stem and seeds are poisonous. Children have been poisoned by eating the berries. The young stems can be forced and blanched and in this condition are not poisonous and can be eaten in place of asparagus in the winter months.

Pokeweed is available from garden centres, or seeds should be sown as soon as ripe. Young plants should be sheltered during the winter and planted out after the danger of frosts is over.

Fruit should be picked when ripe, but care should be used because the juice runs very freely and can stain your clothes. Use the berries in the usual way to produce colours on wool, although these are disappointing if acid is not used; the colours are rather unpleasant dirty creams and yellows. Some recipes advocate using fermented berries, presumably because alcohol is produced, however, because the dye is water-soluble, this seems unnecessary. Adding acid to the dye bath produces an exciting range of reds. Unmordanted wool gives a pale strawberry-pink, alum-mordanted wool gives a deep coral colour, chrome a magenta and copper a pinkish-brown. Adding iron turns the alum-mordanted wool a purple-grey. These colours are vibrant and strong when first dyed, but are disappointingly very light-sensitive and also somewhat sensitive to washing. Exposure to even a few days of sunshine completely discolours the copper-mordanted wool to a pale, dirty orange while the others pale into mere hints of their former glory. Some fading is also noticeable after the skeins have been left in the dark for a few years. The pigment present is thought to be the anthocyanin *cyanidin*.

Al   Cr   Fe   Cu

Safflower

Carthamus tinctorius (Daisy family – Compositae)

Safflower (bastard saffron or dyer's thistle), native to western Asia, has been cultivated around the Mediterranean for many centuries and has become naturalized there and in many parts of south and central Europe and North Africa. It is cultivated for the oil contained in its seeds.

Mummies from Egyptian tombs have been found with bandages still dyed red with safflower and it is mentioned by Pliny in his *Natural History*. It is said to have been introduced into China in the second century BC. Bancroft (1813) classified the red colour as a substantive dye, but he found the red was not permanent in air, sun or with soap. It was used, in his time, particularly on silk to imitate the scarlet of the more expensive cochineal, and the insect dye's rose colour on wool. The red pigment was extracted with soda, precipitated with citric acid, then slowly dried in the shade before being ground into a powder with talc. The pigment was then sold as *rouge végétal*, which was a much safer product than the metallic rouges based on lead which were formerly used. Safflower was used until recently in Britain, where government papers were tied with a cotton tape coloured red with this dye, thus giving rise to the expression 'red tape'. Today, safflower is grown for the oil from the seeds, which is very low in saturated fatty acids. The florets are sometimes sold to unsuspecting tourists in the eastern Mediterranean as saffron. The small sums asked should alert buyers that they are not getting the genuine article at a bargain price.

Safflower is an annual and not a biennial as is recorded in some manuals and seed lists. It can reach 1m (3¼ft) but is usually somewhat shorter. The stems bear unstalked oval or heart-shaped, spiny leaves. The spiny, thistle-like, flower heads have yellow, orange or red tubular florets. The seeds (achenes) are white, sometimes bearing a ring of scales at the apex, but without the crown of hairs characteristic of true thistles.

Safflower is easily grown from seed in any reasonable soil. It is best sown in early spring under glass, then planted out after all danger of frosts is past, leaving 7 to 10cm (3 to 4ins) between plants. Growth is rapid and the flowers can usually be picked from late July onwards. If the red dye is required, flowers should be picked at sundown (in Japan before sunrise is recommended). Fresh flowers usually produce better results than dried, but because a large number is needed to dye even a small quantity of wool, it is necessary to preserve them until you have enough. The florets (not the whole flower head) should be dried quickly and then kept in the dark.

Treating the florets in the usual way gives a yellow dye. Unmordanted wool is dyed a deep, mustard yellow; alum-mordanted wool becomes a paler, but clearer yellow, chrome gives a green-yellow and copper a brassy-gold. The leaves produce pale shades of yellow. To dye red with safflower, the florets should be placed in a muslin or net bag and crushed in the hand or by pounding, then washed in running water until all the yellow, water-soluble dye has been extracted. If your water is alkaline, acidify it slightly, otherwise the red dye will be washed out as well. The red dye is then extracted by soaking in a cold solution of washing soda or other alkali. Bancroft (1813) suggests using just enough alkali to dissolve the red colour then quickly neutralizing with lemon juice, citric acid or mountain ash berries. For unmordanted silk or cotton, soak in the alkali solution then neutralize it with an acid, such as lemon juice. This method of cold dyeing is described by Cardon (1990), who explains that the yellow dye is *carthamin*, a chalcone, which oxidizes to form the red dye *carthamon*. This is present in the orange and red florets and is produced ten hours after the flower opens. It is an unstable substance, which is presumably the reason for the tradition of gathering the florets after dusk or before sunrise.

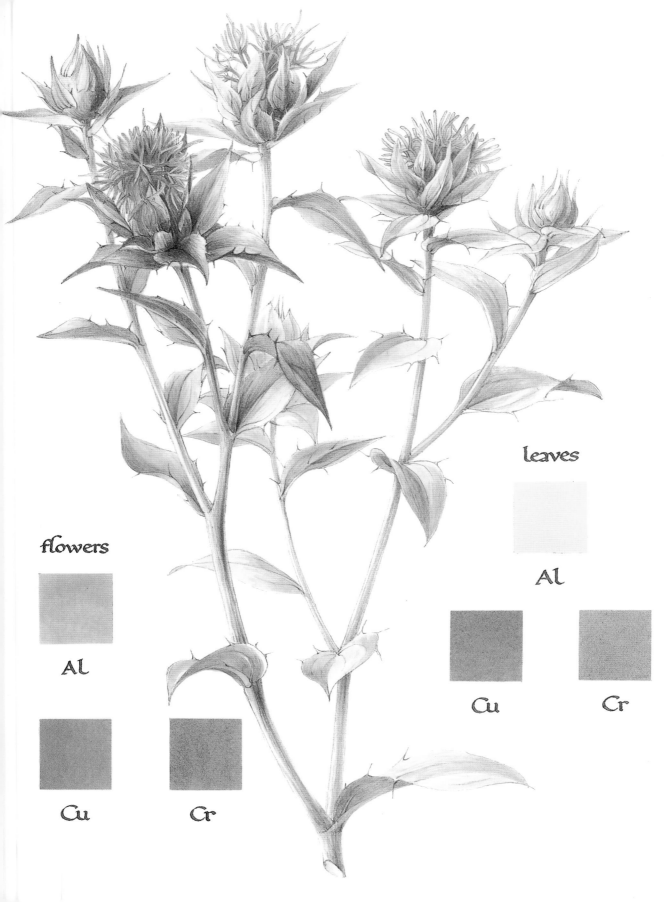

leaves

Al

Cu

Cr

flowers

Al

Cu

Cr

Saffron

Crocus sativus (Iris family – Iridaceae)

Saffron has been cultivated for many centuries and is unknown in the wild. It may persist in a few areas, but it is a sterile hybrid which does not set seed and can only be propagated by dividing the corms. This autumn-flowering crocus (not to be confused with the autumn crocus, *Colchicum autumnale*) has a small corm covered with a finely-netted, fibrous 'tunic'. The 7 to 12 grey-green leaves appear before the flowers in late summer and are 0.5 to 1.5mm (¹⁄₁₆in.) wide. The flowers are lilac-purple, conspicuously lilac-veined and dark stained near the base. The petals are up to 5cm (2ins) long, the anthers yellow and the style red and three-branched.

In the past, saffron was very important because the stigmas produce an excellent substantive dye, needing no mordant to produce its full colour. Saffron was much esteemed in ancient Persia, where documents and the robes of emperors were dyed with it (the saffron robes of Buddhist monks are not dyed with this and the term merely refers to the colour). It is mentioned in the *Song of Solomon* in the Bible, illustrated in Minoan paintings in Crete, and Cleopatra is said to have used it in her cosmetics. It had been introduced into Spain by the conquering Arabs by the twelfth century and, by the late Middle Ages, its cultivation had spread to what are now Italy, France, Switzerland, Austria and Germany. In Basle, in the early fifteenth century, there was a merchant's guild called 'The Saffron'. In England it was so important that the town of Saffron Walden, in Essex, was named after it. It was used as a hair dye in Venice in the late sixteenth century by fashionable ladies. The hair was rubbed with dried caul, egg yolk and honey, and covered with a scarf overnight. The following morning, a large-brimmed straw hat with the crown missing was put on the head, the hair pulled through the hole and laid out over the brim. Saffron and sulphur were then applied to the hair, and the lady would sit on her balcony in full sun throughout the day, while the hair was wetted and dried several times, until the desired colour was achieved (Constantine, 1978).

Saffron corms can be bought from many garden centres and bulb suppliers. It is best grown from fresh, newly divided corms; those sold in garden centres are sometimes very dried out and may disappoint. The corms are best grown in a sandy or loamy soil in a sunny position. In Spain, where it is still grown commercially in the regions of La Mancha and Toledo, the rich, well-drained soil is carefully cultivated and aerated with rakes several times a year. Every four years, the corms are dug up, separated and set out in another plot; the original land is not reused for saffron for another ten years to prevent the build up of pests and diseases and to restore the soil's fertility. The stigmas of the crocus must be harvested every day before the flower wilts. Presumably the dye remains in the stigma of the wilted crocus, but its separation from the petals of the flower would then be much more difficult.

After collection, the stigmas are laid on perforated trays above charcoal fires, a process known as 'roasting'. The procedure of plucking the flowers, separating the stigmas, and roasting them takes place each day until, after about a fortnight, the plants stop producing flowers. Although the amount of dye in one stigma is large in relation to its size, you need a very large number to dye more than a few strands of wool. Packets of the stigmas can be bought from grocers and health-food stores but it is a very expensive commodity and it is called red gold by the Spanish; indeed gram for gram it is nearly as expensive as gold itself.

Colours produced are almost identical on alum-mordanted and unmordanted wool; with chrome the colour is very nearly as bright. They are all very beautiful yellows, while copper-mordanted wool becomes a more greenish-yellow. Pigments present are *crocin* and other carotenoids.

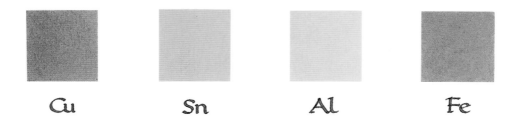

Cu   Sn   Al   Fe

St John's wort

Hypericum perforatum (St John's wort family – Guttiferae)

St John's worts form a genus of about 400 species, many of which are native to Europe. Perforate St John's wort (*Hypericum perforatum*) occurs in Europe, temperate Asia and North Africa and is widely naturalized elsewhere. It is an erect perennial of up to 1m (3¼ft) but usually rather less. The glabrous cylindrical stems have two raised lines along their length. The leaves are unstalked, ovate to linear, and with translucent dots that can be seen when the leaves are held up to the light. The flowers are about 2cm (¾in.) across, in flat-topped inflorescences. The sepals and bright yellow petals have black glandular dots and pointed, somewhat cone-shaped fruiting capsules. It is common in the British Isles, occurring on hedgebanks, in grassland and in waste places, mainly on calcareous soils. There are several other common species of St John's worts with which *H.perforatum* might be confused so a guide to British plants should be consulted.

The black glands of the inflorescence contain a red dye which makes this species of great interest to dyers; the translucent dots in the leaves are sometimes mistakenly thought to be the source of this pigment. Other species may, or may not, have similar dye properties and careful experiments might lead to useful information.

Perforate St John's wort is an aggressive weed which is very poisonous to livestock. Australians have taken great pains to eradicate it using biological control by several species of beetle. Seeds can be bought from specialists; it is an attractive plant and will spread rapidly in any good soil so take care, particularly if you have a small garden.

Flowering tops are used in dyeing and these are traditionally cut at midsummer, especially on St John's day (24 June) after the dew has vanished. This was particularly important if the plant was to be used for medicinal purposes, or for hanging in the doorway to deter witches and evil spirits. Flowering tops picked in August give greenish and brownish colours. Unmordanted wool gives a slightly pinkish-fawn, alum an olive that becomes browner on exposure to sunlight, while chrome-mordanted wool gives a brassy-tan and copper a brownish-tan. Adding iron to alum-mordanted wool gives a deep olive. When we soaked the flowers in alcohol for a short time before dyeing, the colour was a remarkable apple-green with alum, olive-greens with chrome and copper, and yellow with tin; long boiling produced much yellower tones. Various authors record yellows, greens and black; some express disappointment that a red colour, said to be present, is difficult to achieve.

Grierson (1986) reports excellent results only at the height of the flowering season. She obtained a lettuce-green on alum-mordanted wool after boiling for a short time; subsequently, unmordanted wool put into the same dyebath became a maroon colour. A skein left in the dyebath overnight absorbed the remainder of the green and red pigments and was quite black. Yellow was removed by boiling the dye with an alum-mordanted skein. Using this method we achieved a dirty, pale maroon colour, which might have been improved if more yellow pigment had been extracted from the bath with the previous skein.

The yellow pigments are flavonoids, especially *hyperoside* (a *quercetin* derivative), and *rutoside*; carotenoids and tannins are also present. The red dye is *hypericin* (a condensed quinone). This is an unusual pigment which is photosensitive and causes severe skin irritation or even death when farm animals, which have eaten it, are exposed to sunlight. *Hypericum* is said to have been used by dyers in Ecuador to give a yellow colour to their wool, but it is likely to have been a native rather than an introduced species which was involved.

alcoholic
extract

Sn

Al

Cr

Cu

Sanderswood

Pterocarpus santalinus (Pea family – Leguminosae)

Sanderswood (saunderswood, red sandalwood, or redwood) is native to southern India and is extensively cultivated in Madras, Bombay and Bengal. It is a small tree of up to 10m (33ft), with leaves having three leaflets each 10 to 20cm (4 to 8ins) long. It produces few-flowered racemes of yellow flowers and silky pods which are almost 4cm across with a broad wing; the tip of the pod is curved back towards the base.

Sanderswood was known in medieval times, but not necessarily as a dye wood, and its timber is still used today in cabinet making. Marco Polo mentions 'red sanders' growing on two islands east of India. One was probably an island in the Nicobar group, while the other may have been one of the Andaman islands, where *P.indicus* is native. He also gives an account of its importation into China where, historians believe, it was probably used as dye. Its use as a dye wood was certainly known in Europe in the early sixteenth century, and it was sometimes substituted for brazil wood by unscrupulous dealers. Dyers were unable to get any colour because they did not know how to make the dye soluble. Shipments were often of dry pieces of fallen branches and old tree trunks found lying in the forest, and must have been of poor quality. African barwood, and camwood, became increasingly important substitutes at the end of the eighteenth century. Barwood was first exploited by the Portuguese in Sierra Leone, and was imported into Britain in the nineteenth century for cotton printing. The red-brown dye was sometimes called 'mock turkey red'. A red resin from *Pterocarpus*, which exuded from the cut bark and hardened in the air, was known in the eighteenth century as 'dragon's blood', and was valued for its medicinal properties. The red colorant is still used in Africa and India for medicinal purposes and in the latter country for caste marks on the forehead. It is now cultivated on poor soils in areas with a hot, dry climate.

Sanderswood is sold in the form of heartwood chips. They contain a red dye that is insoluble in cold water; stirring the chips in boiling water produces a slightly fluorescent solution, with only a little pigment released. The colour can be readily extracted with alcohol and, to some extent, by strong alkalis such as a solution of 30g (1oz) of washing soda (sodium carbonate) to 100g (3½oz) of fibre. This solution can be used for cotton, but would be harmful to wool. Bancroft (1813) records that the use of sumac, galls or walnut rinds with alkalis partially overcomes the problem of insolubility. He notes that (woollen) broadcloth, pre-mordanted with alum and tartar, and boiled with equal portions of ground sumac and red saunders, gave a very bright and lasting reddish-orange. Wool became almost scarlet when the dye was extracted with alcohol. In our own experiments wool dyed with water-extracted sanderswood produced pale fawns and orange-browns. After soaking for some days, a strong watery extract gave a good range of tans and browns but, extracted with alcohol, the colours were bright copper-orange with no mordant, alum and tin. Chrome gave a deep crimson-tan and copper red-brown. Pigments include flavones, isoflavones, chalcones and pterocarpans, the quinone *santalin*, as well as tannins.

Padouk (or padauk) – *Pterocarpus indicus* – is known as the Andaman redwood, and has very similar dye properties to sanderswood. It is a native of Burma, the Andaman Islands and the Philippines and is a large tree, with leaves of 5 to 9 leaflets, each 10 to 15cm (4 to 6ins) in length. Its dye pigments include the same flavones, chalcones and pterocarpans as sanderswood, but different isoflavones. Camwood, or African sandalwood (*Baphia nitida*), also contains some of the same pigments as *Pterocarpus*, and also has a red dye insoluble in water. All these dyestuffs are known as 'redwoods' and are often substituted for one another.

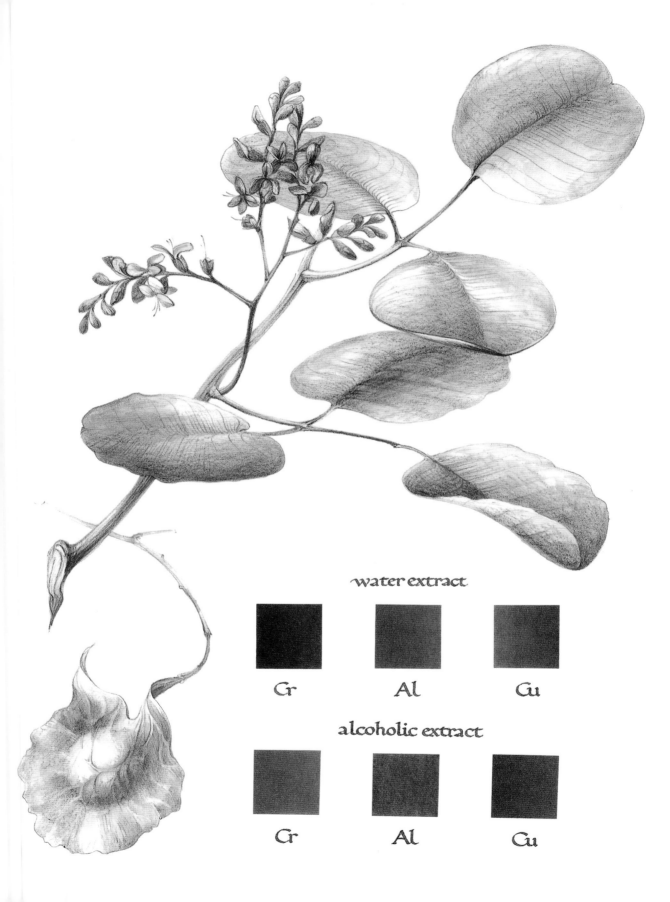

water extract

Cr   Al   Cu

alcoholic extract

Cr   Al   Cu

Saw-wort

Serratula tinctoria (Daisy family – Compositae)

Saw-wort is native to much of Europe, except the extreme north and the Mediterranean and also occurs in North Africa and northern Asia. It is an erect, glabrous, perennial herb of up to 1m (3¼ft) high, but usually rather less. The leaves are often deeply divided, with lobes that are sharply toothed, resembling the teeth of a saw – hence its English name. The thistle-like flower heads are narrowly oblong to ovoid, covered with 6 to 8 rows of scales with short, woolly hairs on the margins; the uppermost scales are often purple. The florets are purple, rarely white, and the seeds (achenes) bear several rows of long bristly hairs. The flower heads may contain either entirely female florets or bisexual and female florets in the same head. The whole plant, except the plant heads, is smooth and hairless.

Little is known of the historical importance of saw-wort, although Linnaeus, the Swedish botanist who founded modern plant classification and nomenclature, refers to its use in eighteenth-century Sweden for dyeing woollen cloth. Bancroft (1813) says it was used to give a durable bright yellow dye with alum and cream of tartar, and a much brighter yellow with tin and cream of tartar.

Saw-wort is now rarely abundant, although it seems to have been more common in the past. It is usually found in copses and light woodlands, or open grasslands on limestone or chalk soils, and can be confused with knapweed (*Centaurea nigra*). This is a similar, but roughly hairy plant with thistle-shaped flower heads, but these are much larger and less cylindrical than those of saw-wort, the florets spread more laterally, the scales are black and comb-like at the top and only a few short hairs are present on the achenes.

Saw-wort can be bought from a few specialist nurseries. The dwarf variant known as *S. seoanei* (sometimes sold under the name *S. shawii*) is often sold by alpine specialists, because it is a low-growing plant that flowers very late in the season. Saw-worts thrive in well-cultivated soil, especially if it contains some lime.

The whole plant, except the roots, is used in dyeing, and it is best picked soon after flowering. It can be used fresh or dried, and is equally recommended for dyeing wool, silk, cotton or linen. Saw-wort gathered in late July gives almost no colour to unmordanted wool, but alum-mordanted wool turns a lemon-yellow. Chrome gives a golden-yellow, copper a yellow-brown, and tin orange-yellow. We produced a beautiful soft green using alum-mordanted wool, top dyed with indigo. The colours are said to be improved when hard tap water is used, or when lime or chalk is added to the dye bath. The pigments present include the flavone *apigenin* and a flavonol – *kaempferol*.

Al

Sn

Cr

Cu

Scabious

Scabiosa columbaria (Teasel family – Dipsacaceae)

The small scabious is a native of Europe, from the Arctic Circle to the Mediterranean, western Asia, Siberia and North Africa. It is found in dry, calcareous pastures, roadsides and on banks. The plant is a perennial with downy stems up to about 70cm (2ft 6ins) but usually much less. It has long-stalked simple or deeply cut leaves at the base, which become shorter stalked towards the top of the stem. The pale lavender flower heads are 1.5 to 3.5cm (½ to 1½ins) across on slender stalks, with a ring of narrow green bracts in a single row below the florets. The outer florets are usually female and rather larger than the inner bisexual florets. All are unequally five-lobed. The ribbed fruit are enclosed in dry cup-shaped structures known as involucels; the persistent calyx forms five bristly lobes on the top of the fruit.

Small scabious is sold by some nurseries and is easy to grow from seed, especially on limy soils. It is perfectly hardy in the British Isles and seed sown in autumn will flower the following summer. Flower heads we collected in August were simmered for half an hour with wool samples. Unmordanted wool and alum-mordanted wool gave very similar bright yellows with a greenish tinge; chrome-mordanted wool gave a warm orange-brown, and copper-mordanted wool gave a yellowish-olive. The colours were all reasonably light-fast except for the copper-mordanted skein which became a little paler on exposure to sunlight. No references have been found in the literature to the use of this species, but we include it because not only does it produce strong, fast colours, but also because it is an attractive garden plant.

Another member of the teasel family used for dyeing is the Devil's bit scabious (*Succisa pratensis*), which has a similar range to that of the small scabious. It has a short rootstock which ends abruptly, giving rise to the superstition that it was bitten off by the Devil. It is found in woodland glades, damp meadows and heaths, preferring a heavy clay soil, often under quite acid conditions. The plant grows up to 1m (3¼ft), the basal leaves are up to 30cm (12ins) long, ovate-elliptic and untoothed. The flower heads are usually deep blue, rarely white or pinkish, and female and bisexual flowers are borne on different plants. The florets are four-lobed and the fruit, surrounded by a four-angled involucel, is crowned by the four short bristles of the calyx.

The basal leaves of Devil's-bit scabious are used for dyeing and are picked before the plant comes into flower. After suitable treatment, similar to that used in woad dyeing (page 114), a yellow liquid containing the pigment *dipsacan* is produced. Fibres soaked in this take up the yellow colour and, when exposed to the air, turn blue by oxidation of the *dipsacan* to *dipsacatin*, so paralleling the reaction of the woad blues. The colour is reputed to be remarkably light-fast. Our own experiments using techniques suitable for woad dyeing were unsuccessful, possibly because the leaves were picked too late in the season. Cardon (1990) refers to *dipsacan* as a pseudoindican, although the compound is not mentioned in the plant biochemistry literature we have been able to consult. The same or similar substances are said to be present in other members of the teasel family, and experiments with the basal leaves of teasels could prove rewarding.

Al

Cu

Cr

Silver birch

Betula pendula (Birch family – Betulaceae)

Silver birch is native to most of Europe, but is confined to mountainous areas in the south; it also occurs in western Siberia, Asia Minor and Morocco. It prefers light sandy soils and often forms dense thickets or woods after forest clearance. This beautiful tree grows up to 25m (82ft), with a single trunk with silver-white, peeling bark, which becomes black and fissured near the base. The purple-brown glossy twigs bear more or less drooping, triangular, toothed leaves. The catkins are pendulous, the male ones being 3 to 6cm (1¼ to 2½ins) long, usually with three minute flowers to each scale, and the female catkins 1 to 3cm (½ to 1¼ins) when in fruit. Tiny winged fruit are shed when the catkins disintegrate in the wind.

The downy birch (*B.pubescens*) is a similar tree. It can be distinguished from silver birch by examining the leaves. Those of silver birch have double-toothed margins, the teeth often curving up towards the apex of the leaf, while those of the downy birch have simply toothed margins, with teeth that do not curve upwards. The twigs of the former often have numerous resin glands; those of the latter lack the glands and are often hairy. *Betula pubescens* has a similar range to *B.pendula*, particularly favouring the poor, acid, peaty soils of moorlands and mountains. Both species are used for dyeing, but no information is available as to their individual properties and they have probably been used indiscriminately by dyers.

Tannins in the bark were used in Scandinavia and Russia for tanning leather. A particularly glandular variety of *B.pubescens* was used in Russia and, after finishing with a distillation of the bark, leather took on a pleasant fragrance. The Swedish botanist Linnaeus referred to the use of the bark for brown dyes in 1749. Dwarf birch (*B.nana*) has a very limited distribution in Britain being confined to mountains in the north, and is said to give a better yellow than silver birch. It is sometimes sold by specialist alpine nurseries as a plant for rock gardens, where it should be planted in a damp spot. Birch leaves have sometimes been used to dye hair, and they are said to produce a slightly greenish-yellow. In the United States and Canada, the yellow birch (*B.alleghaniensis*), gray birch (*B.populifolia*), paper birch (*B.papyrifera*) and the sour or cherry birch (*B.lenta*) are also recorded as being used.

Birches can be bought from garden centres, where the silver birch may be listed under the name *B.alba*. Several varieties are sold, some with deeply toothed or cut leaves, some more strongly pendulous, others even have purple foliage. The latter are said to be weakly growing plants and would presumably contain additional pigments which may or may not be of interest to dyers. Birches grow rapidly in almost any soil and will even grow in chalky soils given a little encouragement. Those with birches in the neighbourhood will probably find seedlings in the garden, which arise only too frequently from the winged fruit.

Leaves, young twigs, young female catkins and the inner bark (not the easily-stripped, papery outer bark) can be used. Bark is said to be best in spring before the leaves open; winter twigs may give a wool a sticky feel, especially if *B.pendula* is used. Leaves picked in July give pale golds and browns when boiled for a short time, while prolonged boiling produces deep oranges and tans. Mature catkins give an attractive range of soft coral and orange-pinks with no mordant, alum and tin. Bark collected from the base of a fallen tree in February gave excellent results. The inner bark, still soft, was used immediately. Alum-mordanted wool gave a strong, dull pink; chrome and copper-mordanted wools gave a light and deep purple respectively. Adding iron to the dye bath gave a soft grey to unmordanted wool, grey-purples with alum and copper and black with purple overtones with chrome. The pigments present are flavonols, derivatives of *hyperoside* and *myricetroside* in the leaves and tannins in the bark.

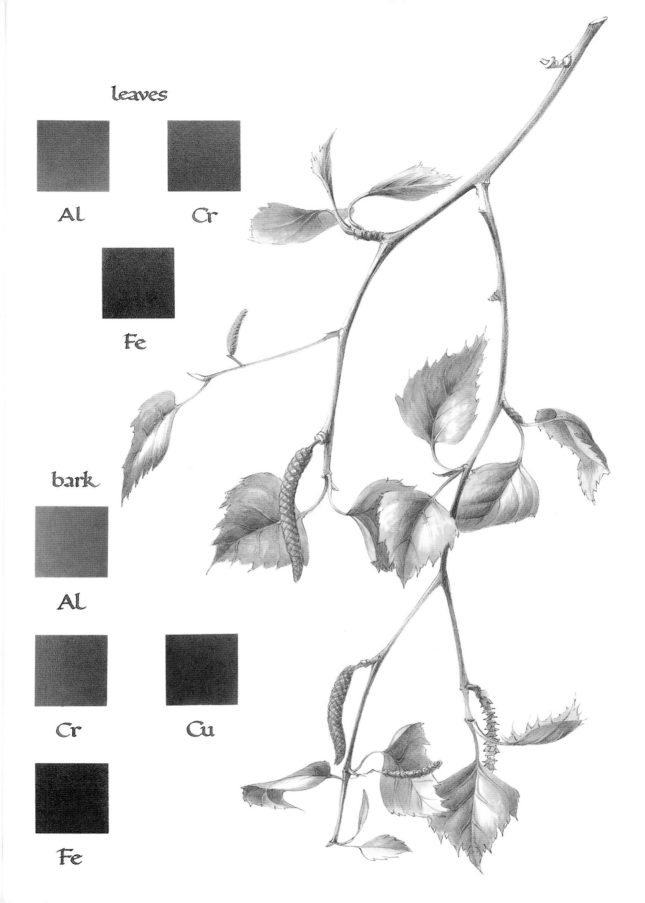

leaves

Al

Cr

Fe

bark

Al

Cr

Cu

Fe

Turmeric

Curcuma longa (Ginger family – Zingiberaceae)

Turmeric, thought to have originated in India, is cultivated throughout the tropics. It has a large fleshy rhizome which is deep yellow and aromatic inside. The long-stalked leaves have oblong blades about 50cm (20ins) long. Flower spikes are about 20cm (8ins) long with conspicuous flowers, the floral bracts are tipped with white, while the sterile bracts are rose coloured; the lips of the flowers are pale to deep yellow.

Turmeric was used in India and the Far East long before it was exported to Europe. Cardon (1990) notes that Marco Polo saw turmeric in Bengal and China. A recipe of 1600 in *Delights for Ladies* by Sir Hugh Platt, advises: 'first wash the hair and by a good warm fire, in warm alum, then with a sponge you may moisten the same in a decoction of turmeric, rhubarb or barberry tree, and so it will receive a most fair and beautiful colour . . . [using] the last water that is drawn from honey skeps being of a deep red colour performeth the same excellently but has a strong smell.' Grierson (1986) records its importation into Scotland in 1612. Bancroft (1813) says that a solution in alcohol was used to dye the yellow spots on the silk bandanas fashionable in England in his time. When washed with soap, the spots became red, but they quickly reverted to yellow when rinsed and dried. Today, it is widely used as one of the ingredients of curry powder and as a natural colouring for mustards, cheeses and cakes, and sometimes as a substitute for saffron. It is also used to colour milk products, paper, wood and wax.

Fresh rhizomes are sometimes sold in large super-markets or oriental food shops. They are difficult to grow from this source and need a warm greenhouse, with a compost of two parts of peat to one of loam. The young roots are soft and very easily damaged by cold, wet soil; they should be grown in slightly damp soil with bottom heat until growth is well established. After flowering, the leaves die back and the rhizomes are then only given sufficient water to keep them from shrivelling until growth resumes in the spring. The colour from fresh rhizomes varies somewhat with the age of the plant and the season in which it is picked. You can also use turmeric powder from the super-market.

Turmeric is a substantive dye, although mordants alter the colour produced. For example, dried rhizomes gave us a lemon-yellow to unmordanted wool. Skeins mordanted with alum, tin, chrome and copper gave a soft orange, orange, pale brownish-orange and a deeper brownish-orange respectively, while iron gave a greenish-orange. Unmordanted acrylic fibre became a deep orange. Another batch of rhizomes gave a greenish tinge with copper-mordanted wool. Cardon (1990) recommends powdered turmeric, but Krohn (1980) says that wool dyed with powder is more light-sensitive than that dyed with rhizomes. However, all the colours are rather fugitive. Cardon states that cotton and silk dyed with turmeric turn red-brown with alkalis, such as soap or even detergents, but these can be neutralized with lemon juice to correct the red tones. The pigments are *curcumin* (known in the food industry as a permitted colour E.100) and other diaroylmethane compounds.

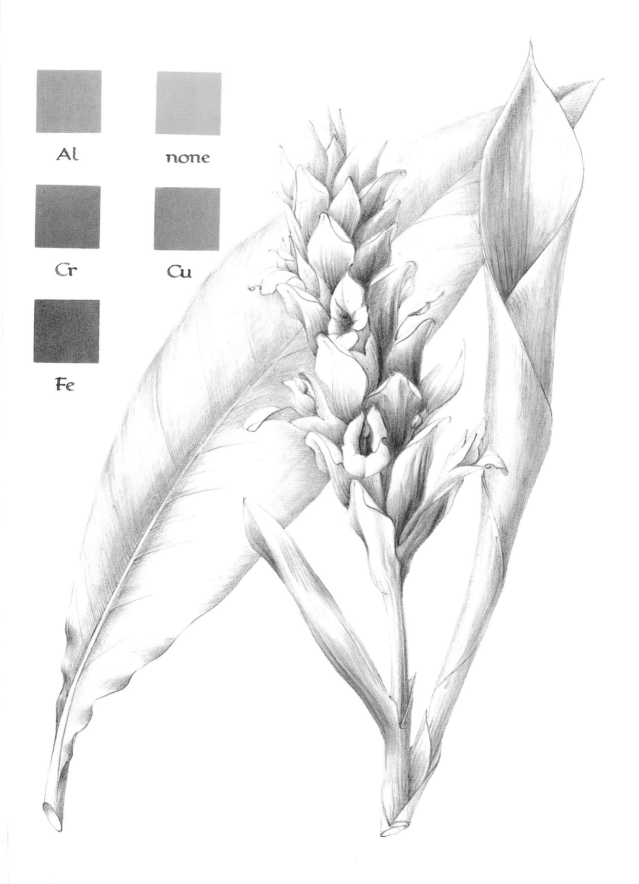

Al

Cr

Cu

Fe

Wallflower

Cheiranthus cheiri (Cabbage family – Cruciferae)

The origin of the familiar cultivated wallflower is unclear and strains which had desirable qualities were probably selected over many centuries. It probably first arose as a hybrid in the Aegean region and is now naturalized on walls, cliffs and rocks throughout central, southern and western Europe, including the British Isles. It is particularly persistent in calcareous habitats. Naturalized plants, introduced into England over 300 years ago, vary in colour and size, with yellow the most frequent petal colour. Wallflowers are perennial, up to 90cm (3ft) high, with narrow lanceolate leaves up to 10cm (4ins) long. The fragrant flowers are borne on stalks and the petals can be up to 2.5cm (1in.) long. The fruit are long and straight, up to 7.5cm (3ins), with a central papery partition, with seeds in a single row on each side.

There are very few references to the use of wallflowers. Bancroft (1813) refers to *Cheiranthus fenestralis* (now known as *Matthiola incana*) as a plant that an aquaintance had told him contained indigo. Cardon (1990) says this plant grows in Crete.

The so-called 'wild' wallflower is available as seed, which should be treated in the same way as the cultivated varieties, but is best planted out in a crack in a wall or in a very dry, chalky or other lime-rich corner. There are many cultivated wallflower varieties, particularly in so far as flower colours are concerned; double and dwarf varieties are also available. Seeds should be planted outdoors in May or June, the seedlings thinned or pricked out, then planted out in their final location in September or October. They make bushy, sturdy plants, which will be covered with blooms in spring, and in a few sheltered localities may even flower at Christmas. Alternatively, plants can be bought in the late summer for planting out. Wallflowers are usually treated as annuals and thrown out after planting. They can, however, be left and will flower again next year, although by this time plants are often straggly and rather unsightly. Blossoms should be handled with care as fingers can become very stained.

Leaves of wallflowers, picked when the old plants were dug up, gave rather muddy, yellow-greens. Mature and faded, dark reddish-brown blossoms gave almost no colour on unmordanted wool, while alum-mordanted wool gave a strong mid-green. Chrome gave a deep olive and copper a soft bluish sage-green. The alum and chrome-mordanted wools became more yellowish when exposed to sunlight, but the copper-mordanted wool merely lightened a little in tone. Dried petals gave a lurid yellow-green with alum, beautiful soft blue-green with no mordant and a greenish-tan with tin.

After picking only a few blossoms, the skin becomes stained blue-black which is extremely hard to remove. It resembles the staining produced by picking a large number of woad leaves. This suggests the presence of a blue pigment in the petals. When treated in the same way as woad leaves (page 114), no blue colour was produced, but vivid yellows and golds resulted with alum, tin and chrome. Flavonols are present including *isorhamnetin*, *robinin* and a derivative of *kaempferol*.

Apart from woad (*Isatis tinctoria*), there are few plants in this large family that have been used for dyeing. Tops of horseradish (*Armoracia rusticana*), and radish (*Raphanus raphanistrum*), produce yellows and browns that are said to be relatively light-fast. Red cabbage (*Brassica oleracea*) is occasionally used for slate-greens, lilacs and purples on wool and silk, but all the colours are extremely sensitive to changes in pH, so great care must be exercised during washing.

Sn

Cu

Al

Al + Sn

Cr

Fe

Weld

Reseda luteola (Mignonette family – Resedaceae)

Weld or dyer's rocket, dyer's mignonette or dyer's weed has sometimes been confused in the literature with dyer's greenweed, due no doubt to the similarity of the vernacular names. It is probably a native of much of south-central and western Europe. Weld has been used as a dye plant for many centuries, and has become naturalized throughout Europe, western Asia, North Africa and has also been introduced into the United States. It is locally common in the British Isles, where it is usually found on calcareous soils along roadsides, on dry banks and in waste places, particularly where the soil has been disturbed.

Seeds of weld have been found in Neolithic lake villages in Switzerland, where it is assumed to have been used for its dye. It is said to have been used by the Romans for dyeing wedding garments and the robes of the vestal virgins (Wickens in Buchanan (ed.) 1990). Pliny says its use was exclusively for women's garments. In 1243, according to Cardon (1990), the *Capitolaribus de Tinctorum* of Venice, the oldest rules of the dyer's profession in the West, forbade the use of other yellows, particularly that from dyer's greenweed, because weld was thought to give much more permanent colours. It was used as one of the dyes to colour the yellow caps the Jews of Europe were compelled to wear in the Middle Ages. In England, it was used with dyer's greenweed, together with woad or indigo, to produce the colour known as Lincoln or Kendal green. Weld was gradually superseded by black oak which, weight for weight, produced a greater strength of colour.

Weld is a biennial herb of up to 1.5m (5ft) and produces in its first season a rosette of narrow leaves up to 8cm (3¼ins) long, with undulating margins. In the second season, it produces flowering spikes bearing smaller, entire or divided leaves. The spike is initially single, becoming branched as the plant matures. The very numerous yellow-green flowers give rise to small capsules, which open before they become ripe; the numerous green seeds then gradually become black. The inflorescences are remarkable because they always point towards the sun, following it during the day from east to west. The Swedish botanist Linnaeus noted that it pointed north at midnight – perhaps in response to the midnight sun in the far north.

Weld is best grown from seed and, once established, it will resow itself readily, sometimes in unwanted places if the seed harvest is delayed. Seeds are available from specialist seed suppliers and should be sown in autumn where the plants are to grow to maturity. Seeds sown in February or March, when the soil is warm, may flower as annuals late in summer.

Whole plants, except the roots, may be used fresh or dried and should be hung up in bundles to dry in a warm place. Some authors think that fresh material can give slightly brighter colours, but it was used commercially in the dried state, as whole plants, or chopped or powdered. The yellows and olives produced by weld are very light-fast. Unmordanted wool takes up scarcely any dye at all; wool mordanted with alum or tin produces bright yellows, chrome an orange and copper a yellow-olive. In fade tests, alum-mordanted wool became much more orange and the copper-mordanted wool faded a little, while the chromed wool did not alter significantly. When iron was added to the dye bath, chrome-mordanted wool became a pale yellow-olive and alum-mordanted wool a deep olive. The quantity of dye in the plants and the colours produced vary considerably from season to season, and sometimes the addition of soda may be helpful (Grierson, 1986). She also says that iron, when added to the dye bath, caused precipitation of the dye. Cardon (1990) advocates using small plants, before branching takes place, for the best results. The pigments include the flavones, *luteolin* and *apigenin*. The former seems to be specific to weld; the closely related wild mignonette (*R.lutea*) does not contain this pigment.

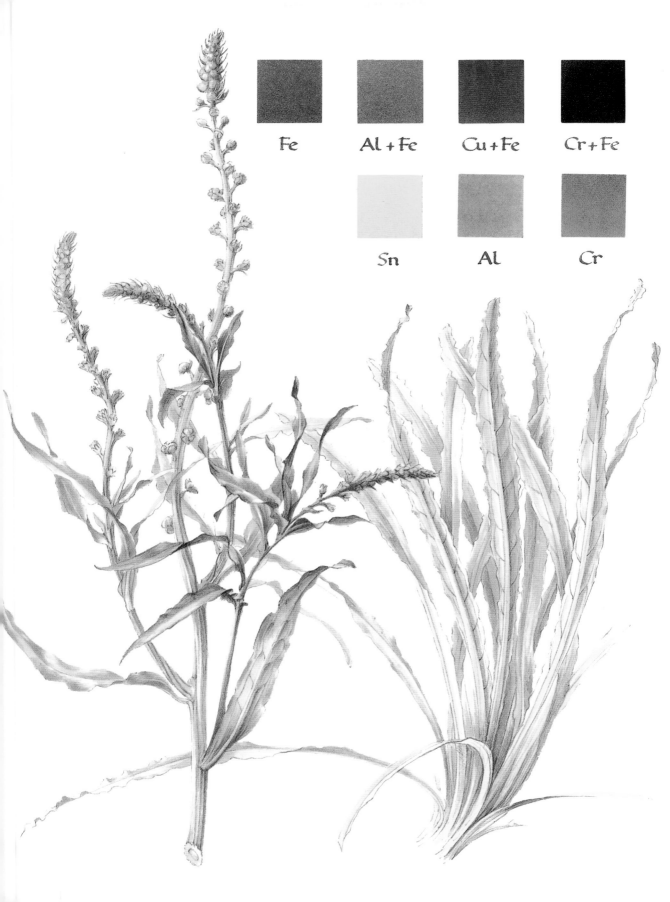

Fe Al + Fe Cu + Fe Cr + Fe

Sn Al Cr

White water-lily

Nymphaea alba (Water-lily family – Nymphaeaceae)

The white water-lily is native throughout most of Europe and it also occurs in western Asia and parts of India. In the British Isles it is found in the sheltered parts of lakes, ponds, slow-moving rivers and canals and is tolerant of low nutrient levels. The plant has a stout, fleshy, creeping rhizome which bears translucent submerged leaves, long-stalked floating leaves and in shallow water, emergent leaves up to 30cm (12ins) wide. The floating flowers are slightly fragrant as they open and are up to 20cm (8ins) in diameter. Four green sepals cover the egg-shaped buds and the numerous white petals gradually become narrower towards the centre and grade into the bright-yellow stamens. The fruit ripens under water, shedding the many seeds which float for a few minutes and then sink into the mud. Flowering occurs from June till late autumn.

The white water-lily was one of the most important sources of black for traditional wool dyeing in Scotland. Cardon (1990) says it was used commercially around Nuremberg in the last century, where it was also used for grey, black and violet dyeing on wool and for textile printing.

Rhizomes of water lilies are available from garden centres and nurseries specializing in aquatics. The rhizomes should be planted in water at least half a metre deep, in weighted nets or plastic baskets to prevent floating. Several cultivars of *N.alba* are sold, including plants with pink or purplish flowers. Wild plants are extremely fast growing and will cover a small pond very quickly. *Nymphaea candida* is a related smaller species with many garden varieties which are better for smaller ponds. They probably have similar dye properties to the white water-lily, but we have not tested them and no literature references have been found.

Rhizomes should be gathered when the water-lily is divided or thinned-out. In the wild, they were traditionally collected any time in the summer, when the lake or pond level fell sufficiently for the gatherers to wade into the water conveniently. Rhizomes cleaned, chopped and simmered for an hour produce browns on mordanted wool and a pale fawn on unmordanted wool. Much darker browns result if iron is added to the dye bath and the wool simmered for a further fifteen minutes. All these colours are reasonably light-fast. Grierson (1986) obtained blue-grey black from roots that were simmered for two days with acid and double quantities of iron. She also reports a honey-olive colour from the leaves. Cardon (1990) advocates crushing the rhizomes with a wooden stick or pestle; when macerated in a blender the rhizome blackens the stainless-steel blade, which is then quite difficult to clean. Pigments present are tannins including *gallotannins*, *ellagitannins* and condensed tannins.

Al   Cu   Cu+Fe

Woad

Isatis tinctoria (Cabbage family – Cruciferae)

Woad is native to Europe, western Asia and North Africa, and has been cultivated in Europe for many centuries. It is rare in England, but quite common in south and central Europe. It prefers light calcareous or sandy soils and is a very attractive plant both in flower and fruit. As a biennial (or perennial) herb, it forms a rosette of bluish-green, strap-shaped leaves during the first season of growth. Flowering stems up to 1.5m (5ft) are produced the following year, bearing smaller stem-clasping leaves. The inflorescence is spreading, many branched, with numerous yellow flowers; the fruit are black and pendulous.

There is a huge amount of literature relating to the history of woad and this probably exceeds the information about any other dye, except indigo. Egyptian hieroglyphics of 1500 BC refer to the use of blue dyes. There is some disagreement whether this was woad or indigo (Cardon, 1990), and the author also mentions recent excavations in France of sites in Neolithic caves where woad was certainly used. It is also known that Celtic tribes used woad as a skin dye and the name of Britain is said to derive from the Celtic word *brith*, which means paint. Julius Caesar in 54 BC referred to Britons who, without exception, stained themselves with woad, giving them a wild look in battle. Glastonbury's name is said to derive from *glastum* or blue (woad); Somerset was a centre of the woad industry. By the Middle Ages woad had become very important throughout Europe. The buildings where it was crushed and fermented were known as woad mills and the foul smell from them was such that Elizabeth I forbade its production within five miles of any of her estates. It was gradually superseded by indigo, but not before laws were passed throughout most of Europe to prevent the import of the foreign dye and protect the local woad growers. The classic historical account of woad by Hurry (1930) refers to its use for dyeing police uniform material.

Woad is very easy to grow from seed and plants can be bought from some nurseries, but transplanting from the container can damage the tender root. Seeds are available from specialists, or from dyers because the plant produces abundant fruit. It is classed as a noxious weed in some parts of the United States.

The flowering tops can be used to give yellowish dyes, but the desirable blue dye comes from the leaves, especially those of the first year's growth. They should be picked immediately before use and are unsatisfactory if frozen or dried. If the largest leaves are plucked, leaving 4 to 5 young leaves, a further crop will be ready after a few weeks. If they cannot be used at once, leaves should be crushed and kneaded into balls of at least tennis-ball size and dried in a draught of air for several weeks. Goodwin (1982) gives an excellent account of woad cultivation and use. Woad 'balling' was the means by which it was preserved in the past, and was almost exactly similar to that described for *Polygonum tinctorium* (page 70).

The dye can be extracted by several methods: a watery extract can be made from fresh or preserved leaves by fermentation with, or without, adding urine, but always in an alkaline medium; or it can be extracted chemically by methods similar to those used for indigo. The best method is to pour nearly boiling soft water over chopped leaves. After half an hour, the leaves are squeezed and thrown away and the liquid treated with alkali – caustic soda is convenient. The extract is then heated to about 50°C (122°F) and sodium dithionite is added to a water bath at the same temperature. The two are then combined and the pre-soaked wool is dyed by successive dippings and airings as for indigo. The colour depends on the number of dippings and airings, the source of the plants, the season, and is little altered by mordants. The pigment, *isatan B*, is similar to that of indigo, and is hydrolyzed into *indoxyl* (leuco-indigo); two molecules of this then form one molecule of *indigotin* (indigo blue) when oxidized. *Indirubin* is also present.

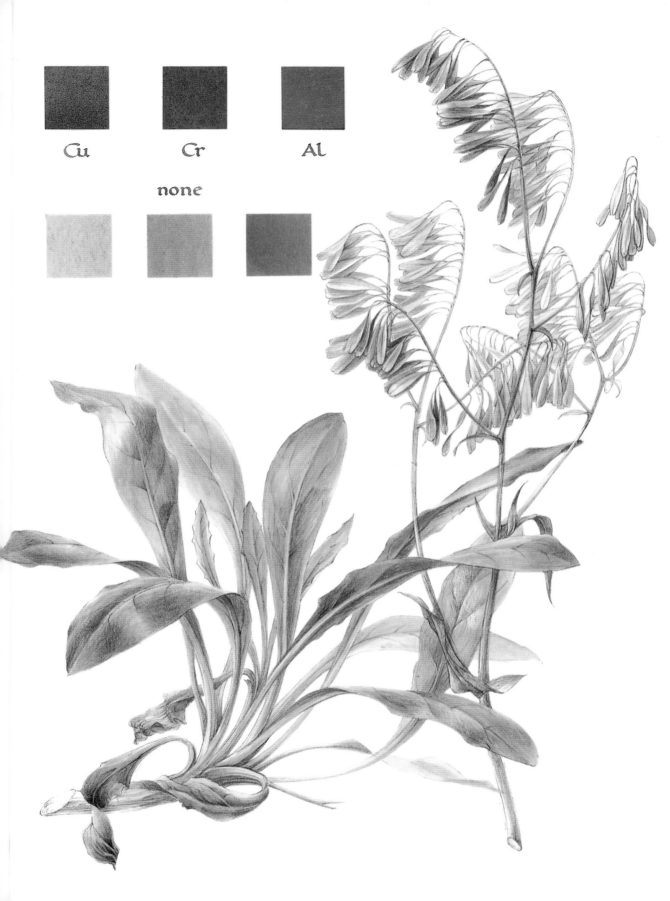

Cu   Cr   Al

Yew

Taxus baccata (Yew family – Taxaceae)

Yew is native to Europe, Asia as far as the Himalayas, and North Africa, being confined to the mountains towards the southernmost parts of its range. In the British Isles it is a lowland species, particularly of chalk and limestone, and sometimes forms almost pure woods on sheltered places on the chalk of southeast England. It is a large, very long-lived, coniferous tree, with a broad trunk and a thin, reddish-brown bark. The green twigs are covered with numerous dark green leaves of 1 to 3cm (½ to 1¼ins), in two ranks, one on each side. Male and female flowers occur on different trees, the males in small cones which produce copious yellow pollen, while the female flowers are solitary. In fruit, the seed is surrounded by a fleshy, cup-like, red aril which ripens in August or September and is attractive to birds.

Most of the references to the use of yew are from this century by modern craft dyers. The trees are so slow growing, and so highly regarded by craftsmen, that it seems unlikely material would be available in sufficient quantities for commercial dyeing.

Yews are available from most garden centres and many cultivars are offered, including prostrate and very upright forms, with leaves of various shades of gold or even bluish-black, and arils of different colours. It is a splendid hedging plant, but very slow growing. It takes many hundreds of years to reach the size of the magnificent old trees found in English churchyards, some of which are said to be more than a thousand years old. Yew can be grown on almost any well-cultivated soil, even light acid ones – although it prefers some lime. It is tolerant of deep shade and some of the prostrate forms are recommended as ground cover in shady places.

The heartwood is the part most commonly used for dyeing. It can be obtained as chips from dye suppliers, or perhaps as waste from craftsmen, who use it for expensive but beautiful furniture and carvings. The heartwood gives soft oranges and tans which become slightly deeper with acid, while alkali seems to have no effect. Various authors report browns, pinks or orange-reds, depending on the methods used. The leaves may also be used and give a soft tan with alum-mordanted wool and beautiful browns with chrome and copper. The bark gives a variety of shades of pinkish-fawns and browns with different mordants. The red arils are of very little use; a chamois colour after very lengthy boiling with alum has been reported. Pigments present in the leaves include carotenoids of which *rhodoxanthin* is one (known in the European food industry as E.161(f)); this and other carotenoids and anthocyanins are found in the fruits and other parts of the plant; *hydroxycyanidin* is recorded from the heartwood.

wood   leaves

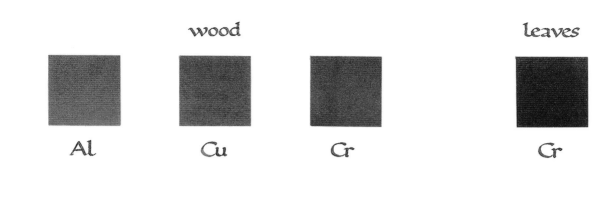

Al   Cu   Cr   Cr

Young fustic

Cotinus coggygria (Sumac Family – Anacardiaceae)

The smoke tree, or Venetian sumac, the source of the dye called young fustic, is native to southern and central Europe, the Middle East, India and China and is usually found on dry, rocky slopes. It forms a bush or small, much-branched tree of up to 5m (16ft), with stalked, rounded leaves. The inflorescences at the end of the branches are composed of a few flowers, borne on much-branched, very slender, hairy stalks, together with numerous sterile stalks. In the autumn, the inflorescences dry and turn smoky grey, giving the impression of a cloud of smoke.

The smoke tree is a highly regarded garden shrub and is available from most garden centres and might be listed under the old synonym *Rhus continus* or, rarely, *Cotinus obovatus*. The latter is a related American species which has brilliant autumn colours and *C.coggygria* plants are sometimes named as this species. The true *C.coggygria* has fawn-coloured inflorescences but several cultivars have purple-coloured sprays and others purple leaves. The best variety for autumn colour seems to be 'Flame'; 'Royal Purple' is also popular. They all grow readily in many soils. The purple forms are very disappointing in positions where the sun does not shine through the leaves, and flowering is inhibited by shade. It can be propagated by layering; the young branches are pegged down and covered with soil, into which new roots emerge. It is also possible to take cuttings.

The wood of the smoke tree was traditionally used for dyeing. It was described as producing yellow-orange, browns or yellows of poor colour-fastness, often being used with other yellow dyes, such as old fustic, to heighten colours. Experiments with bark from an unknown variety of *C.coggygria* with both green and purple-tinged leaves gave interesting results. The bark was stripped from the twigs and the outer, very thin, flaky bark was separated from the inner green layers. The inner bark gave pale fawns when used with various mordanted wools, while the outer bark gave brilliant tones: pinkish on unmordanted wool, strong orange-tan with alum, glowing gingery-tan with chrome and pale chocolate with copper. All these colours were moderately light-fast. Pigments present are flavonols, particularly *fisetin* and *myricetin*; an aurone, *sulphuretin*, is also present, as are tannins.

Young fustic was used in Europe during the Middle Ages and was still of great importance in the economy of poor Provençal villages early last century. The original name for this species was *Rhus cotinus* and this has been a source of confusion for non-botanists today, many of whom have thought it to be the same as tanner's or Sicilian sumac (*Rhus coriaria*). Tanner's sumac has leaves which are almost evergreen and have winged stalks, and fruits that are brownish purple. It resembles the familiar, crimson-fruited, stag-horn sumac (*R.typhina*) of our gardens, a related species from eastern North America.

Tanner's sumac is found on rough, rocky hillsides, where it grows into a small tree. Its leaves are used to dye wool and silk, and for cotton printing. Bancroft (1813) says that bark from the branches, trunk and roots of tanner's sumac give fugitive yellows and oranges. The bark is used with other dye materials, for example, with madder to produce the colour known as turkey red, with sanderswood to give various browns, and with logwood and iron to give black. Tanner's sumac was of considerable importance in the Middle Ages, particularly for its tanning properties. Bancroft notes its use to give a very pale yellow with alum, but a good black with iron. He also states that the berries of smooth sumac (*R.glabra*) were used as a mordant by North American Indians when dyeing porcupine quills with roots of bedstraws (*Galium* spp).

Many other species of *Rhus* contain considerable amounts of dye substances, especially tannins, in their leaves. Stag-horn sumac (*R.typhina*) and several other species from North America, including shining sumac (*R.copallina*), are recorded as producing browns from leaves.

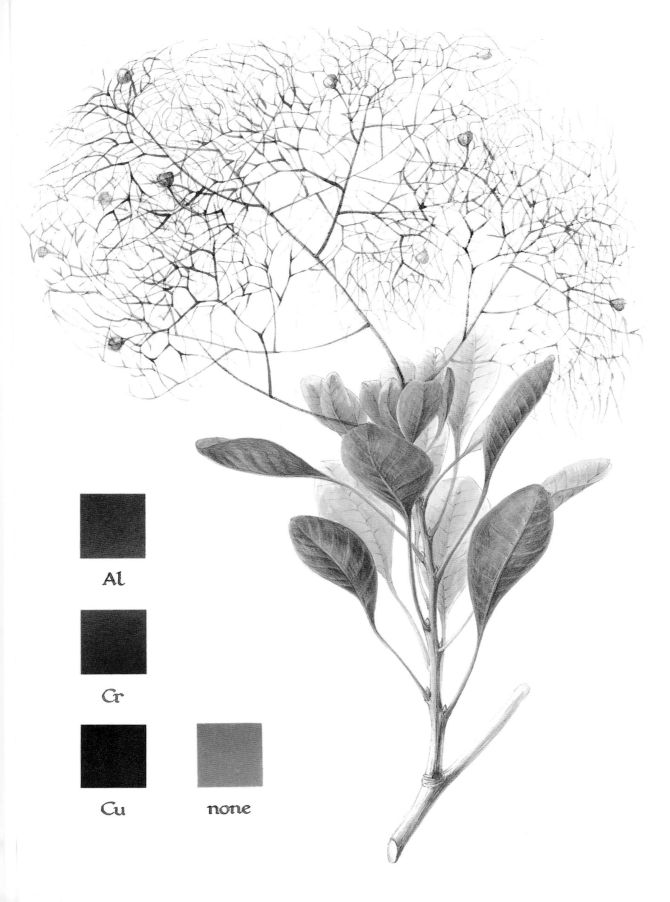

Al

Cr

Cu   none

Useful addresses

AUSTRALIA AND NEW ZEALAND

HANDWEAVERS' GUILD INC., P.O. BOX 24090, Auckland, New Zealand

SOUTH AFRICA

HANDWEAVERS' GUILD, c/o St Paul's Road, Houghton, 2196 Johannesburg

UNITED KINGDOM

THE ASSOCIATION OF THE GUILDS OF WEAVERS, SPINNERS AND DYERS, Anne Dixon, 2 Bower Mount Place, Maidstone, Kent

GEORGE WEIL/FIBRECRAFTS, Old Portsmouth Road, Peasmarch, Guildford GU7 2QD www.fibrecrafts.com. [*Fibres, books, dyes, mordants*]

HANDWEAVERS STUDIO & GALLERY LTD, 29 Haroldstone Road, London E17 7AN www.handweaversstudio.co.uk [*Fibres, books*]

HILLTOP SPINNERS AND WEAVERS CENTRE, Windmill Cross, Canterbury Road, Lyminge, Folkestone, Kent CT18 8HD [*Fibres, dyes*]

P & M WOOLCRAFTS, Pindon End, Hanslope, Milton Keynes MK19 7HN www.naturaldyes.co.uk [*Fibres, books, dyes, mordants, seeds*]

SUFFOLK HERBS LTD, Sawyers Farm, Little Cornard, Sudbury, Suffolk CO10 ONY [*seeds*]

UNITED STATES AND CANADA

EARTH GUILD, 33 Haywood Street, Asheville, North Carolina 28801, USA www.earthguild.com

HANDWEAVERS GUILD OF AMERICA, INC, Two Executive Concourse, Suite 201, 3327 Duluth, Duluth, Georgia 30096, USA www.weavespindye.org

HANDWEAVERS, SPINNERS AND DYERS OF ALBERTA, c/o Birgit Rasmussen, President, 7202 112 Street, Edmonton, Alberta T6G 1J3, Canada www.hwsda.org

THE MANNINGS HANDWEAVING SCHOOL, P.O. Box 687, East Berlin, Pennsylvania 17316, USA www.the-mannings.com

ONTARIO HANDWEAVERS AND SPINNERS, 2097 Gary Crescent, Burlington, Ontario L7R 1T1, Canada www.ohs.on.ca

RIO GRANDE WEAVER'S SUPPLY, 216-B Pueblo Norte, Taos, New Mexico 87571, USA www.weavingsouth-west.com

RUMPELSTILTSKIN, 1021 R Street, Sacramento, California 95814, USA

SUSAN'S FIBER SHOP, N250 Highway A, Columbus, Wisconsin 53925, USA www.susansfibershop.com

WEAVERS' AND SPINNERS' GUILD, West Point Grey Community Centre, 4397 West Second Avenue, Vancouver, British Columbia, Canada V6R 1K4 www.anwg.org/bc/vancouver

Reading list

ADROSKO, R. J. 1971. *Natural Dyes and Home Dyeing.* pp. 154. New York, Toronto and London. Dover

AKEROYD, J. R. & STEARN, W. T. 1985. Joice Nankivell Loch (1893–1982), *Annales Musei Goulandris* 7. pp. 17–24

BAINES, P. 1989. *Linen, Hand Spinning and Weaving.* pp. 208. London. Batsford

BANCROFT, E. 1813. *The Philosophy of Permanent Colours,* Vols 1 & 2. pp. 542. London. Cadell & Davies

BLAMEY, M., FITTER, R. & FITTER, A. 1977. *The Wild Flowers of Britain and Northern Europe.* pp. 208. London, etc. Collins

BUCHANAN, R. (Ed.) 1990. *Dyes from Nature.* pp. 96. Reprinted from *Plants and Gardens – Brooklyn Botanic Garden Record* **46**. Brooklyn. Brooklyn Botanic Garden

BUC'HOZ, P. J. 1800. *Manuel Tinctorial des Plantes* (ed. 5). pp. 290 Paris. Buc'hoz

CARDON, D. 1990. *Guide des Teintures Naturelles.* pp. 340. Neuchâtel and Paris. Delachaux and Niestlé.

CLAPHAM, A. R., TUTIN, T. G. & WARBURG, E. F. 1981. *Excursion Flora of the British Isles.* pp. 499. Cambridge, London, New York, etc., Cambridge University Press

CONSTANTINE, M. 1978. *Herbal Hair Colouring.* pp. 38. London. The Herb Society

DALBY, G. 1985. *Natural Dyes, Fast or Fugitive.* pp. 48. Minehead. Ashill Publications

DALBY, G. & CHRISTMAS, L. 1984. *Spinning and Dyeing – an introductory manual.* pp. 135. Newton Abbot, London and North Pomfret, Vermont. David and Charles

DONY, J. G., JURY, S. L. & PERRING, F. 1986. *English Names of Wild Flowers.* (ed. 2). pp. 117. London. Botanical Society of the British Isles

DUNSMORE, S. 1988a. Growing Nettles in Nepal. *Journal for Weavers, Spinners and Dyers.* no. 145. pp. 9–10

DUNSMORE, S. 1988b. Weaving Nettle in Nepal. *Journal for Weavers, Spinners and Dyers.* no. 146. pp. 17–18

GOODWIN, J. 1982. *A Dyer's Manual.* pp. 128. London, New York etc. Pelham Books, Stephen Greene Press (Penguin Group)

GRIERSON, S. 1986. *The Colour Cauldron.* pp. 234. (privately published), Tibbermore, Perth. Grierson

GRIERSON, S. 1989. Dyeing Distinctions. *Journal for Weavers, Spinners and Dyers.* no. 150. pp. 20.

GRIERSON, S. 1989. *Dyeing and Dyestuffs.* pp. 32. Princes Risborough. Shire Publications

HECHT, A. 1989. *The Art of the Loom – Weaving, Spinning and Dyeing across the World.* pp. 208. London. British Museum Publications

HURRY, J. B. 1930. *The Woad Plant and its Dye.* pp. 328. London. Oxford University Press

KROHN, V. F. 1980. *Hawaii Dye Plants.* pp. 136. Honolulu. University Press of Hawaii

MAIRET, E. M. 1916. *Vegetable Dyes* (ed. 1). pp. 45. London. Faber & Faber

PHILIP, C. 1994. *The Plant Finder* (ed. 8). Whitbourne. Headmain Ltd. For The Hardy Plant Society. [Valuable list of nurseries offering unusual plants].

POMET, P. 1694. *Histoire Général des Drogues.* Paris. Publisher unknown

PONTING, K. G. 1981. *A Dictionary of Dyes and Dyeing.* pp. 207. London. Bell & Hyman

ROBERTSON, S. M. 1973. *Dyes from Plants.* pp. 144. New York. Van Nostrand Reinhold

SCHETKY, E. McD. (ed.) 1964. *Handbook on Dye Plants and Dyeing* pp. 101. Reprinted from *Plants and Gardens – Brooklyn Botanic Garden Record.* **20**. Brooklyn. Brooklyn Botanic Garden

STACE, C. 1991. *New Flora of the British Isles.* pp. 1226. Cambridge etc. Cambridge University Press

WATT, G. 1889–93. *Dictionary of the Economic Products of India.* 6 volumes. London. Allen

WICKENS, H. 1983. *Natural Dyes for Spinners and Weavers.* pp. 96. London. Batsford

WEIGLE, P. (ed.) 1973. *Natural Plant Dyeing.* pp. 65. Reprinted from *Plants and Gardens – Brooklyn Botanic Gardens Record.* **29**. Brooklyn. Brooklyn Botanic Garden

Checklist of scientific names

Arranged by families

ANCARDIACEAE
Cotinus coggygria Scop. [*Rhus cotinus* L.]
C.obovatus Raf.
Rhus copallina L.
R.coriaria L.
R.glabra L.
R.typhina L.

ARALIACEAE
Hedera helix L.

BERBERIDACEAE
Berberis asiatica DC.
B.dumetorum Gouan
B.laurina Thunb.
B.thunbergii DC.
B.vulgaris L.
Mahonia aquifolium (Pursh.) Nutt.
M.japonica (Thunb.) DC. [*M.beali* Carr.]
M.lomariifolia Takeda

BETULACEAE
Betula alleghaniensis Britton
B.lenta L.
B.nana L.
B.papyrifera Marsh.
B.pendula Roth
B.populifolia Marsh.
B.pubescens Ehrh.

BIXACEAE
Bixa orellana L.

BORAGINACEAE
Alkanna tinctoria (L.) Tausch
Anchusa virginica L.
Lithospermum arvense L.
L.caroliniense (J.F. Gmel.) MacM.
L.incisum Lehm.
Onosmodium virginicum (L.) A.DC.
Pentaglottis sempervirens (L.) Tausch

CAPRIFOLIACEAE
Sambucus ebulus L.
S.nigra L.

COMPOSITAE
Anthemis arvensis L.
A.cotula L.
A.tinctoria L.
Carthamus tinctorius L.
Centaurea nigra L.
Chamaemelum nobile (L.) All.
Chrysanthemum segetum L.
Coreopsis tinctoria Nutt.
Cosmos sulphureus Cav.
Dahlia pinnata Cav.
Matricaria discordea DC.
Serratula seoanei Willk. [*S.shawii* Hort.]
S.tinctoria L.
Solidago canadensis L.
S.flexicaulis L.
S.graminifolia (L.) Salisb.
S.virgaurea L.
Tagetes erecta L.
T.micrantha
T.patula L.
Tripleurospermum inodorum (L.) Schultz Bip.

CRUCIFERAE
Armoracia rusticana P. Gaertner, B. Mayer & Scherb.
Brassica oleracea L.
Cheiranthus cheiri L. [*Erysimum cheiri* (L.) Crantz]
Isatis tinctoria L.
Matthiola incana (L.) R.Br. in Aiton [*Cheiranthus fenestralis* L.]
Raphanus raphanistrum L.

DIPSACACEAE
Scabiosa columbaria L.
Succisa pratensis L.

ERICACEAE
Calluna vulgaris (L.) Hull
Erica arborea L.
E.herbacea L. [*E.carnea* L.]
E.cinerea L.
E.mediterranea L.
E.tetralix L.

FAGACEAE
Quercus petraea (Mattuschka) Liebl.
Q.robur L.
Q.velutina Lam. [*Q.tinctoria* Bart.]

GUTTIFERAE
Hypericum maculatum Crantz
H.perforatum L.

IRIDACEAE
Crocus sativus L.

JUGLANDACEAE
Carya illinoensis (Wangenh.) K.Koch
Juglans cinerea L.
J.nigra L.
J.regia L.

LEGUMINOSAE
Acacia catechu (L.f.) Willd.
Baphia nitida Afzel. ex Lodd.
Baptisia australis (L.) R.Br.
B.tinctoria (L.) Vent.
Caesalpinia bahamensis Lam.
C.bicolor C.H.Wright
C.brasiliensis L.
C.crista L.
C.echinata Lam.
C.sappan L.
Chamaespartium saggitale (L.) P.Gibbs
Cytisus scoparius (L.) Link [*Sarothamnus scoparius* (L.) Wimm. ex Koch]
Genista anglica L.
G.pilosa L.
G.tinctoria L.
Haematoxylon brasiletto Karsten
H.campechianum L.
Indigofera arrecta Hochst. ex A. Rich.
I.argentea L.
I.heterantha Brandis [*I.gerardiana* R.C. Grah.]
I.suffruticosa Mill.
I.tinctoria L.
Peltophorum brasilense (Sw.) Urban
Pterocarpus indicus Willd.

P.santalinus L.f.
Spartium junceum L.

LILIACEAE
Allium cepa L.
A.oschaninii B.Fedsch.
A.schoenoprasum L.
A.ursinum L.
Colchicum autumnale L.

LYTHRACEAE
Lawsonia inermis L.

MELIACEAE
Toona ciliata M.Roehm.

MORACEAE
Broussonetia papyrifera (L.) Vent.
Chlorophora tinctoria (L.) Benth.
Maclura pomifera (Raf.) Schneid.
Morus alba L.

MYRTACEAE
Eucalyptus bicostata Maiden, Blakley
& J.Simm.
E.cinerea F. Muell. ex Benth.
E.cordata Labill.
E.crenulata Blakley & Beuzev.
E.globulus Labill.
E.gunnii Hook.f.
E.macrorhyncha F. Muell. ex Benth.
E.nicholii Maiden & Blakley

NYMPHAEACEAE
Nymphaea alba L.

N.candida C.Presl

PALMACEAE
Areca catechu (L.f.) Willd.

PAPAVERACEAE
Sanguinaria canadensis L.

PHYTOLACCACEAE
Phytolacca americana L.

POLYGONACEAE
Fagopyrum esculentum Moench
Polygonum aviculare L.
P.tinctorium Aiton
Fallopia japonica (Houtt.) Ronse
Decraene [*Polygonum cuspidatum*
Sieb. & Zucc.]

RESEDACEAE
Reseda lutea L.
R.luteola L.

RHAMNACEAE
Rhamnus alaternus L.
R.catharticus L.
R.dahurica Pall.
R.saxatilis Jacq. [*R.infectorius* L.]
R.utilis Decne.

ROSACEAE
Rubus allegheniensis Porter
R.fruticosus L.
R.idaeus L.
R.tricolor Focke

RUBIACEAE
Galium aparine L.
G.mollugo L.
G.odoratum (L.) Scop.
G.saxatile L.
G.verum L.
Morinda citrifolia L.
Rubia cordifolia L.
R.peregrina L.
R.tinctoria L.
Uncaria gambier Roxb.

SALICACEAE
Salix alba L.
S.caprea L.
S.cinerea L.
S.fragilis L.
S.purpurea L.
S.viminalis L.

TAXACEAE
Taxus baccata L.

URTICACEAE
Boehmeria nivea (L.) Gaudisch.
Girardinia diversifolia (Link)
I.Friis
Urtica crenulata Sw.
U.dioica L.
U.urens L.

ZINGIBERACEAE
Curcuma longa L. [*C.domestica*
Valeton]

Botanical glossary

ACHENE A dry fruit, which does not split open, containing one seed

ADVENTITIOUS ROOTS Roots other than those of the original root system, for example, roots arising from climbing stems

ANTHER The part of a stamen containing the pollen

ARIL A fleshy outgrowth from a seed

BIPINNATE See under PINNATE

BISEXUAL Having male and female organs in the same flower

BRACTS Small leaf-like structures closely associated with flowers

CALCAREOUS Derived from chalk and limestone rocks

CALYX The sepals forming a group which surrounds the petals and protects the flower bud

CAPSULE A dry fruit that splits open by various means to release the seeds

CATKIN A spike of minute flowers adapted for wind pollination

COMPOSITE FLOWERS A head of minute florets like those of a daisy or dandelion

COMPOUND LEAF A leaf composed of several distinct leaflets

CORM A tuberous rootstock covered with thin papery scales

CULTIVAR A cultivated variety not found in the wild

DISK FLORET The small tubular flowers which form the centre of a composite flower, for example, the daisy

DECIDUOUS Woody perennials which shed their leaves before the winter

ELLIPTIC A leaf broadest in the middle and narrow at both ends

FLORA The assemblage of plants living in an area. Flora with a capital F means a technical manual for plant identification

GENUS A group of closely related plant species, for example, buttercups

GLABROUS A smooth surface without hairs or other projections

GLOBOSE Spherical, globe-shaped

HERMAPHRODITE Having both male and female reproductive organs

HUSK The soft, green outer covering of a fruit such as a walnut or horse chestnut

INFLORESCENCE An assemblage of flowers grouped together

INVOLUCEL Small secondary bracts

LANCEOLATE Shaped like a spear blade

LEAFLET A leaf-like segment of a compound leaf

LINEAR Narrow and parallel sided

NODE A point on a stem from which one or more leaves arise

NUT A dry, one-seeded fruit with a hard outer shell

NUTLETS Tiny, dry fruit, like miniature nuts

OBCORDATE Heart-shaped with the point nearest the stem

OBLONG Parallel-sided but broader than linear

OBOVATE Egg-shaped with the narrower end nearest the stem

OBTUSE Blunt

ORBICULAR Round in outline

OVATE Egg-shaped with the broader end nearest the stem

PALMATE A compound leaf having four or more leaflets arising from a single point

PANICLE A much-branched inflorescence

pH A measure of acidity or alkalinity. Low values are acid and high ones are alkaline; pH 7 is neutral

PINNATE A compound leaf in which the leaflets are arranged in two rows on either side of a midrib. The primary divisions of bipinnate leaves are themselves pinnate

PROSTRATE Lying on, or very close to the ground

RACEME A spike of stalked flowers

RACHIS The axis of a compound leaf bearing the leaflets

RAY FLORETS The petal-like outer florets of a composite flower, for example, a daisy

RECEPTACLE The base of a flower on which the petals and other floral parts are borne

RECURVED With the tip curved back

RHIZOME An underground, rooting stem

RHOMBOID Diamond-shaped

ROSETTE A group of leaves radiating from a stem at soil level

SCALE A minute flat projection from a surface

SEPAL A member of the outermost (usually green) ring of flower parts

SESSILE Without a stalk

SPIKE An inflorescence of stalkless flowers on an undivided stem

STAMEN A male (pollen-producing) organ in a flower

STIGMA The female organ of a flower which receives the pollen

SYNONYM A name given to a plant already named by an earlier author

TRIFOLIATE Having three leaflets

TUBER A thickened, underground root or stem

TUNIC The papery outer coverings of a bulb or corm

UMBEL An inflorescence with branches arising from one point like the spokes of an umbrella

UNISEXUAL Flowers with either male or female organs

WHORL Leaves or other plant organs arranged in a ring

Index